10th EPS-QEOD Europhoton Conference on Solid-State, Fibre, and Waveguide Coherent Light Sources (EUROPHOTON 2022)

EPJ Web of Conferences Volume 267 (2022)

Hannover, Germany
28 August – 2 September 2022

Editor:

P. Helfenstein

ISBN: 978-1-7138-6267-3

Printed from e-media with permission by:

Curran Associates, Inc.
57 Morehouse Lane
Red Hook, NY 12571

Some format issues inherent in the e-media version may also appear in this print version.

For additional information, please contact EDP Sciences – Web of Conferences at the address below.

EDP Sciences – Web of Conferences
17, Avenue du Hoggar
Parc d'Activité de Courtabœuf
BP 112
F-91944 Les Ulis Cedex A
France

Phone: +33 (0) 1 69 18 75 75
Fax: +33 (0) 1 69 28 84 91

contact-edps@webofconferences.org

Additional copies of this publication are available from:

Curran Associates, Inc.
57 Morehouse Lane
Red Hook, NY 12571 USA
Phone: 845-758-0400
Fax: 845-758-2633
Email: curran@proceedings.com
Web: www.proceedings.com

TABLE OF CONTENTS

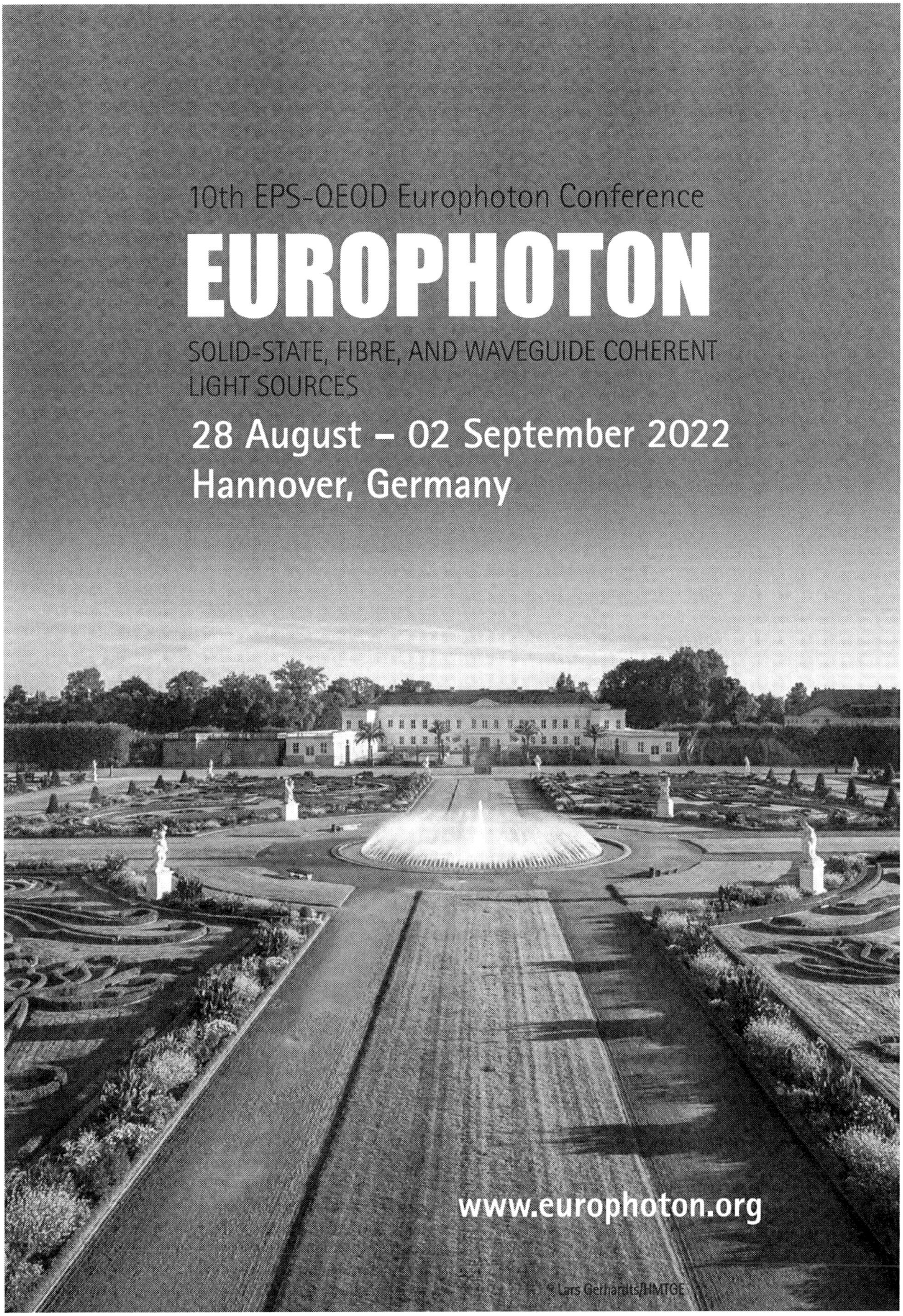
10th EPS-QEOD Europhoton Conference
EUROPHOTON
SOLID-STATE, FIBRE, AND WAVEGUIDE COHERENT LIGHT SOURCES
28 August – 02 September 2022
Hannover, Germany
www.europhoton.org
© Lars Gerhardts/HMTGE

10th EPS-QEOD Europhoton Conference
EUROPHOTON
SOLID-STATE, FIBRE, AND WAVEGUIDE COHERENT LIGHT SOURCES
Program includes
Two-day Summer School
Half-day special Symposium
General Chair
Uwe Morgner, Cluster of Excellence PhoenixD, Leibniz University Hannover, Germany
Programme Chair
Markus Pollnau, Advanced Technology Institute, University of Surrey, United Kingdom
Sub-Committee Chairs
Jacob Mackenzie, University of Southampton, United Kingdom
Andrejus Michailovas, EKSPLA, Lithuania
Summer School Speakers:
Marc Hanna, CNRS - Université Paris-Saclay, France
Ursula Keller, ETH Zürich, Switzerland
Michael Kues, Leibniz University Hannover, Germany
Anne L'Huillier, University of Lund, Sweden
Derryck Reid, Heriot-Watt University, Edinburgh, United Kingdom
Andreas Tünnermann, University and Fraunhofer-Institute Jena, Germany
For more information visit:
www.europhoton.org
For exhibit opportunities contact:
conferences@eps.org
This conference is organized by the European Physical Society in cooperation with the Quantum Electronics and Optics Division (QEOD) of the European Physical Society (EPS).
Conference venue:
Hannover, Germany
Abstract submission deadline:
Thursday, 2nd June 2022
Leibniz Universität Hannover
PhoenixD
Photonics · Optics · Engineering
Innovation Across Disciplines
EPS
in cooperation with
VolkswagenStiftung
European Physical Society
6 rue des Frères Lumière
68200 Mulhouse • France
www.eps.org
Photo: Lars Gerhardts/HMTGE

edp sciences

EPJ Web of Conferences

Statement of Peer review

In submitting conference proceedings to EPJ Web of Conferences, I certify to the Publisher that I adhere to the **Policy on Publishing Integrity** of the journal in order to safeguard good scientific practice in publishing.

1. All articles have been subjected to peer review administered by the proceedings editors.
2. Reviews have been conducted by expert referees, who have been requested to provide unbiased and constructive comments aimed, whenever possible, at improving the work.
3. Proceedings editors have taken all reasonable steps to ensure the quality of the materials they publish and their decision to accept or reject a paper for publication has been based only on the merits of the work and the relevance to the journal.

Title, date and place of the conference

10TH EPS-QEOD EUROPHOTON CONFERENCE ON SOLID-STATE, FIBRE, AND WAVE GUIDE COHERENT LIGHT SOURCES
28 AUGUST 2 SEPTEMBER 2022, HANNOVER, GERMANY

Proceedings editor(s):

HELFENSTEIN EUROPEAN PHYSICAL SOCIETY

Date and editor's signature OCTOBER 4TH, 2022

EUROPEAN PHYSICAL SOCIETY
6 rue des Frères Lumière
F-68200 ULHOUSE
Tél. 03 8 32 94 40

EUROPEAN PHYSICAL SOCIETY EPS MULHOUSE

 https://doi.org/10.1051/epjconf/202226701001

LED-pumped CTH:YAG luminescent concentrator as broadband incoherent source in the SWIR

Lisa Lopez, Pierre Pichon, Frédéric Druon, Patrick Georges and François Balembois
Université Paris-Saclay, Institut d'Optique Graduate School, Centre National de la Recherche Scientifique, Laboratoire Charles Fabry, 91127, Palaiseau, France

CTH:YAG is a laser medium know for many years [1]. The atmospheric transmission properties and eye safe nature of this laser emitting at 2.1 µm make it attractive for many applications. Thanks to efficient energy transfer between Cr^{3+}, Tm^{3+} and Ho^{3+}, CTH:YAG has often been pumped in the visible by flashlamps [2]. CTH:YAG has also a remarkable broadband spectrum spanning from 1.7 µm to 2.15 µm. Sensing and inspection applications are growing with the emergence of low-cost image sensors in the SWIR range. Illumination sources in this wavelength band are limited: halogen, black body and low power LEDs. The purpose of this work is to show that CTH:YAG opens opportunities for innovative, high-brightness, spectrally-broadband and incoherent sources in the SWIR.

As pump source, we used a Ce:YAG luminescent concentrator (LC) [3], since CTH:YAG absorption band matches the Ce:YAG yellow emission. A 1×14×220 mm³ Ce:YAG is pumped by 2240 blue LEDs (Fig.1a). A 1×3×14 mm³ CTH:YAG is bonded on the output surface by an UV-curing optical adhesive. Total internal reflections are preserved by the index difference between CTH:YAG (n=1.82) and the adhesive (n=1.5). The CTH:YAG sample operates as a LC with a maximum brightness on the smallest face (1×3mm²) chosen as the output face. A gold mirror is positioned on the opposite face to increase the output power.

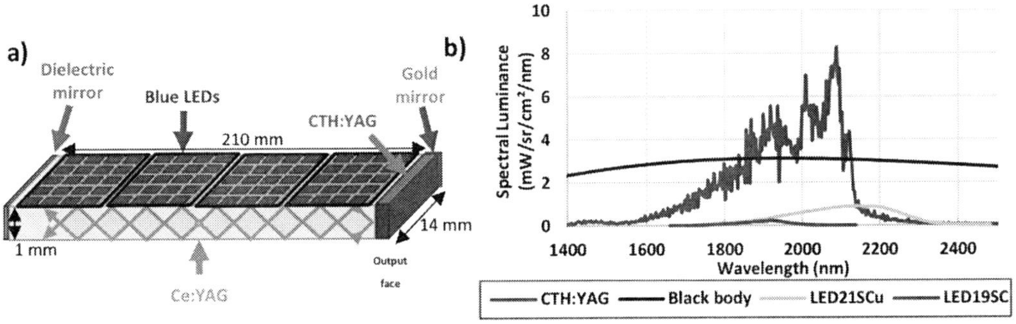

Figure 1: a) Scheme of the cascade of concentrators with blue LEDs pumping Ce:YAG that pumps CTH:YAG
b) Spectral brightness of a blackbody at T=1500 K, two infrared LEDs manufactured by Boston electronics and our CTH:YAG concentrator. (The black body operates continuously, the other sources in QCW.)

LEDs operate in quasi-continuous wave with 260 µs pulses at 10 Hz and a peak power of 3.15 W per LED. The Ce:YAG delivers a peak power of 1,7 kW towards the CTH:YAG. The spontaneous emission from the output face is collected by an integrating sphere photodiode power sensor (S148C Thorlabs). We measured a SWIR peak power of 50 mW on a 1.6 to 2.1 spectrum (Fig1 b). Despite a low efficiency, the spectral brightness is 3 times higher than the one of a black body at T=1500 K and approximately 10 times higher than LEDs operating in this wavelength range. Many improvements are possible regarding crystal cooling, light extraction and pump duration (as CTH:YAG lifetime is 8.5 ms). This first result proves that a LED pumped cascade of concentrators has a potential for alternative SWIR sources with applications like active imaging.

[1] B. M. Antipenko, V. A. Buchenkov, A. S. Glebov, T. I. Kiseleva, A. A. Nikitichev, and V. A. Pismennyi, "Spectroscopy of YAG :CrTmHo laser crystals," Opt. Spectrosc. (USSR),**64**, pp. 772-774, (1988).
[2] Dan Bar-Joseph "2.097µ Cth:YAG flashlamp pumped high energy high efficiency laser operation (patent pending)", Solid State Lasers XXVII: Technology and Devices, Proc. SPIE **10511**, 1051107 (2018)
[3] P. Pichon, A. Barbet, J-P. Blanchot, F. Druon, F. Balembois, and P. Georges, "LED-pumped alexandrite laser oscillator and amplifier", Optics Letters, **42**, pp.4191-4194 (2017).

Multi-point, pulse-train laser ignition of methane-air mixtures by a high-peak power passively Q-switched Nd:YAG/Cr^{4+}:YAG compact laser

Nicolae-Tiberius Vasile, Gabriela Croitoru, Ciprian Dumitrache, Nicolaie Pavel
National Institute for Laser, Plasma and Radiation Physics, Laboratory of Solid-State Quantum Electronics, Magurele 077125, Romania
nicolaie.pavel@inflpr.ro

Laser ignition (LI) of various fuel mixtures is a research topic that seeks to address mankind's concern for the environment and natural resources through a number of specific features [1]. In the case of LI there is no electrode placed in the combustion chamber and therefore the flame kernel develops freely. Moreover, using proper focusing optics, LI can be initiated at any point in the engine cylinder, thus providing a possibility to optimize engine performance. Furthermore LI is a suitable technique for igniting lean fuel mixtures, thus reducing emissions without decreasing engine power [2-4]. A laser beam offers also the possibility to control the ignition spatially, by simultaneous ignition in several points [5,6], as well as temporally, by using a train of laser pulses for a single ignition event. In the case of LI at several points, the combustion time is reduced and the flame can lose less heat, resulting in higher temperatures and higher pressures. All experiments conducted so far on LI at several points have used large commercial lasers, and the laser beams have been directed to and then focused in a static combustion chamber or in the engine cylinder using various optical systems [6]. In addition, there are very few reports of multi-pulse laser ignition [7]. In this work, we report on the realization of a compact, diode-pumped passively Q-switched Nd:YAG/Cr^{4+}:YAG laser with four beams, operating in single-pulse as well as in burst mode (yielding trains of up to five pulses). This device was used to study LI of CH_4-air mixtures in a combustion chamber, employing spatial ignition control as well as temporal ignition control mode.

A composite Nd:YAG/Cr^{4+}:YAG ceramic medium (1.1-at.% Nd) was used to build the laser, considering a monolithic resonator design. The Nd:YAG/Cr^{4+}:YAG laser was pumped with four fiber-coupled laser diodes; the pump pulse duration was chosen so that each laser beam contains trains of up to five pulses at a set repetition rate. The energy of an individual laser pulse was 3.2 mJ with duration of 0.9 ns. LI of CH_4-air mixtures was done in a spherical combustion chamber, properly equipped for such experiments [8].

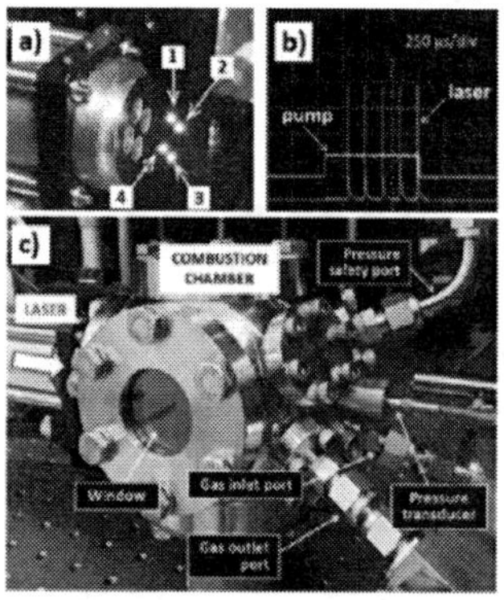

Fig. 1 a) The laser with four points of ignition and **b)** a train of five pulses is shown. **c)** The combustion chamber is presented.

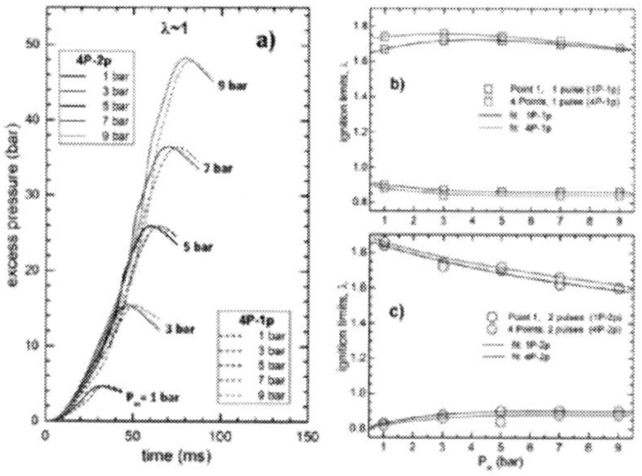

Fig. 2 a) Pressure vs. time for LI in four points of CH_4-air stoichiometric mixtures (λ~1) at various initial pressure, P_{in}. Limits of ignition for λ, considering ignition with **a)** a laser pulse and **b)** trains of two laser pulses.

LI of CH_4-air mixtures with different values of relative equivalence ratio, λ was investigated considering four ignition schemes: ignition in point 1 with one laser pulse (1P-1p) and with train of two laser pulses (1P-2p); four-point ignition with one laser pulse (4P-1p) and with two laser pulses (4P-2p). Maximum pressures and the combustion times were obtained from the pressure curves recorded during LI (Fig. 2a); an analysis is made for each ignition scheme, considering initial pressures in the combustion chamber up to 9 bar. Ignition limits for λ have been determined (Fig. 2b, Fig. 2c) and a discussion of these results is presented. The advantages of multi-point LI with trains of laser pulses are analyzed compared to LI at one location. Multi-point LI with burst-mode train pulses is a feasible technique for igniting lean fuel-air mixtures and for increasing the combustion speed.

Financing Projects PN-III-P4-PCE-2021-0519 and NUCLEU-LAPLAS V 16N/2019, Ministry of Research and Innovation, Romania.

[1] P.D. Rooney, Opt. Eng. **33**(2), 510-521 (1994).
[2] J. D. Dale, P. R. Smy, and R. M. Clements, SAE Technical Paper 780329 (1978).
[3] T. Taira, S. Morishima, K. Kanehara, N. Taguchi, A. Sugiura, M. Tsunekane, "World first laser ignited gasoline engine vehicle," in: 1st Laser Ignition Conference, Yokohama, Japan, 2013 paper LIC3-1.
[4] N. Pavel, M. Bärwinkel, P. Heinz, D. Brüggemann, G. Dearden, G. Croitoru, O. V. Grigore, Prog. Quantum Electron. **58**, 1-32 (2018).
[5] T. Phuoc, Combust. Flame **122**(4), 508-510 (2000).
[6] P. Patane, M. Nandgaonkar, Opt. Laser Technol. **130**, 106305 (2020).
[7] S. Lorenz, M. Bärwinkel, R. Stäglich, W. Mühlbauer, D. Brüggemann, International Journal of Engine Research **17**(1), 139-150 (2016).
[8] N.-T. Vasile, N. Pavel, Opt. & Laser Techn. **141**, 107169 (2021).

EPJ Web of Conferences **267**, 01003 (2022) https://doi.org/10.1051/epjconf/202226701003
EUROPHOTON 2022

Iterative 3D modeling of thermal effects in end-pumped continuous-wave Ho^{3+}:YAG lasers

Marius Rupp[1, 2], Katharina Goth[1, 2], Marc Eichhorn[1, 2], and Christelle Kieleck[1]

[1] *Fraunhofer IOSB (Institute of Optronics, System Technologies and Image Exploitation), Gutleuthausstraße 1, 76275 Ettlingen, Germany*
[2] *Institute of Control Systems, Karlsruhe Institute of Technology, Fritz-Haber-Weg 1, 76131 Karlsruhe, Germany*

Corresponding author: marius.rupp@iosb-extern.fraunhofer.de

High power lasers operating at around 2 µm can be used for a variety of applications, ranging from medical treatments and countermeasures to the pumping of mid infrared optical parametric oscillators (OPOs). Especially for the latter application, pulsed crystal lasers have shown great potential. Ho^{3+}-doped media like Ho^{3+}:YAG are efficient for emitting at wavelengths greater 2 µm, where OPOs can be pumped more efficiently due to lower parasitic absorption. To optimize Ho^{3+}:YAG resonator setups for high power, efficiency and a good M^2, a simulation model can greatly assist with the development and understanding of such setups. The model we present in this work aims to simulate a real resonator setup with high accuracy, therefore many properties of the resonator and the laser gain medium need to be taken into consideration.

The fundamental propagation model is based on a split-step fast Fourier transform beam propagation method, similar to the one developed by Fleck et al. [1]. By adding additional calculation modules in between the propagation steps, a resonator with a variety of components can be simulated. For example, mirror surface components are necessary to calculate the replicating field in a resonator. Another essential component of the model is the laser crystal, the gain of which is calculated by solving the temporal rate equations of the material, in this case Ho^{3+}:YAG. The rate equations are based on a 2D model for Er^{3+}:YAG described by Eichhorn [2] and include upconversion, cross-relaxation and quenching effects to model the laser gain as accurately as possible. The laser crystal component also takes into account thermal effects like thermal lensing and stress-induced birefringence. This is achieved by iteratively calculating the temperature distribution and the thermal expansion in the crystal based on finite-difference methods solving the respective differential equations [3].

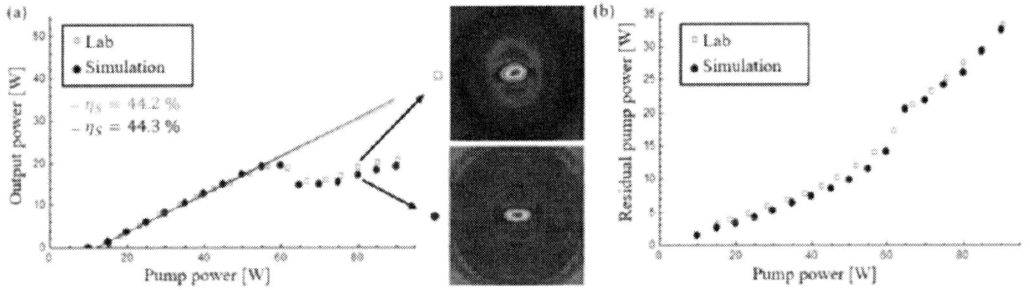

Figure 1. Comparison of an experimental Ho^{3+}:YAG resonator setup with the model. In (a), the laser output power is compared, in (b) the residual pump power.

The combination of the aforementioned algorithms allows the modeling of Ho^{3+}:YAG resonator setups with an accurate representation of laser output power, resulting M^2 and output field distribution. Fig. 1 shows a comparison of the model with a simple single pass, Tm^{3+} fiber end-pumped Z-resonator that was not optimized in order to test the model at suboptimal beam qualities due to thermal distortions. The output Ho^{3+}:YAG laser power (a) as well as the residual pump power after the single pass through the crystal (b) is in excellent agreement throughout the whole pumping range. The model is also capable of reproducing the beam distortions in the output field with very good accuracy, which is shown in Fig. 1 for a pump power of 80 W. The validation with this resonator setup proves the model to be a powerful tool to investigate and optimize Ho^{3+}:YAG resonators, which will be put to the test and improved upon in further research work.

References

[1] Fleck Jr, J. A.; Morris, J. R.; Feit, M. D., "Time-Dependent Propagation of High Energy Laser Beams through the Atmosphere", Appl. Phys. **10**, 129-160 (1976)
[2] Eichhorn, M., "Numerical Modeling of Diode-End-Pumped High-Power Er^{3+}:YAG Lasers", IEEE J. of Quant. El. **44**(9), 803-810 (2008)
[3] Cousins, A. K., "Temperature and Thermal Stress Scaling In Finite-Length End-Pumped Laser Rods", IEEE J. of Quant. El. **28**(4), 1057-1069 (1992)

Passively Q-switched Er:YAP laser generating 21 ns pulses at 2.9 μm

Richard Švejkar*, Dominika Popelová, Jan Šulc, Helena Jelínková

Czech Technical University in Prague, Faculty of Nuclear Sciences and Physical Engineering, Břehová 7, Prague, Czech Republic, 11519
**richard.svejkar@fjfi.cvut.cz*

Laser sources emitting in the mid-infrared spectral range from 2 to 8 μm can be utilized in various applications (spectroscopy, remote sensing, polymer processing, medicine, etc.). [1,2]. Laser generation in short-wavelength part of this range (around 3 μm) can be directly obtained from diode-pumped erbium-doped active. For medical and spectroscopy applications, there is particularly a desire for compact systems generating short pulses with high peak power in this spectral range. From this point of view, the passive or active Q-switching method is an ideal method for pulse generation. Passive Q-switching with saturable absorber allows building a very compact laser resonator without requirements on additional power sources or special crystals. Moreover it is a simple method to obtain the short pulses close to 3 μm. [1-3] The goal of this contribution is investigation of Er:YAP laser emitting radiation in a Q-switching regime with the shortest possible pulses.

The tested crystal Er:YAP (5 at. % of Er^{3+}, b – cut, Crytur) had a form of a rectangular block (9 mm in length) with polished faces (3 × 3 mm). The Er:YAP was placed inside the hemispherical laser resonator (length 45 mm) consisting of a flat pumping mirror and spherical output coupler (OC, radius 50 mm, R = 95 % @ 2.9 μm). The laser diode (emitted wavelength 976 nm, focusing optic 1:2, and waist beam diameter 250 μm) operated in pulsed regime, repetition rate 10 Hz and pulse duration 5 ms, was used as pumping source. The passive Q-switching was realized by semiconductor saturable absorber (SA-2800-10-10ps-5.0-12.7g-e, Batop GmbH). The energy probe (J10-MB-LE, Coherent) was utilized for the measurement of the output energy of the Q-switched pulses. The pulse duration was measured using a photodiode (PVI-4TE-6, time constant 0.5 ns, VIGO System S.A.) connected to an oscilloscope (TDS3052B, 500 MHz, 5GS/s, Tektronix).

Firstly, the Er:YAP laser was tested in a free-running regime without a saturable absorber to characterize the output laser parameters. In a free-running laser regime, the slope efficiency, maximum output mean power, and laser threshold were reached with respect to the absorbed mean power of 25.5 %, 200 mW, and 67 mW, respectively. Er:YAP laser emitted radiation at 2920 nm in fundamental mode. Consequently, the SA was placed into the resonator within Er:YAP and OC, then the Q-switching laser was successfully realized. In Fig. 1(a) the measured pulse duration (under maximal pumping) is presented. The shortest output pulses reached 21.8 ns at FWHM with peak power with a corresponding peak power of 24.8 W, pulse energy of 0.54 μJ, and a repetition rate of 41.6 kHz. The pulse duration and repetition rate were decreased with decreasing pumping power which is presented in Fig. 1(b). This type of compact laser system can be further utilized in spectroscopy, alternatively in polymer processing if the pulse will be amplified to a sufficient energy level. We believe that using a saturable absorber with higher absorption should further increase the pulse energy and peak power.

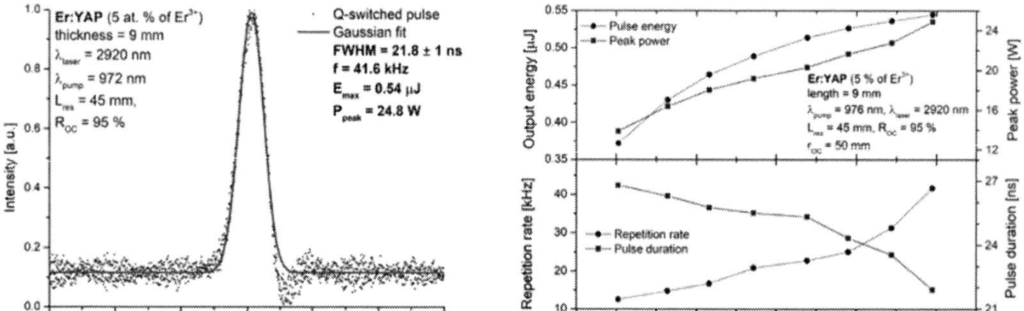

Fig. 1 Short pulse generated in Q-swtichitng (a) and characterization of Q-switching regime (b); Δt – pumping pulse duration, f – frequency, λ_{pump} – pumping wavelength, λ_{laser} – emission wavelength, FWHM – pulseduration, P_{peak} – peak power, E_{max} – maximal pulse

References

[1] Rüdiger Paschotta, *Encyclopedia of Laser Physics and Technology*, 1. Edition (Wiley-VCH, Berlin, 2008).
[2] Švejkar, R., Šulc, J., Jelínková, H., Er-doped crystalline active media for ~ 3 μm diode-pumped lasers, Progress in Quantum Electronics, 2020, vol. 74, p.100276, ISSN 0079-6727, doi:10.1016/j.pquantelec.2020.100276.
[3] J. Frauchiger, W. Lüthy, P. Albers, H. P. Weber, Laser properties of selectively excited $YAlO_3$:Er, Opt. Lett. 13 (11) (1988) 964–966. doi:10.1364/OL.13.000964.

0.5 - 1.3 GHz tunable pulse repetition rate solid state laser generating 230 fs pulses with 200 mW average power

Roger Wüst[1], Daniel Hug[1], Benjamin Rudin[2], Florian Emaury[2], Bojan Resan [1,3*]

1. Institute of Product and Production Engineering, University of Applied Sciences and Arts Northwestern Switzerland (FHNW), Klosterzelgstrasse 2, Windisch, CH-5210, Switzerland

2. Menhir Photonics, Industriestrasse 42, Glattbrugg, CH-8152, Switzerland

3. Faculty of Medicine, Josip Juraj Strossmayer University, Josipa Huttlera 4, Osijek, 31000, Croatia

**bojan.resan@fhnw.ch*

In the recent years the demand for ultrafast laser systems with various parameters in terms of wavelength, power, and pulse repetition rate has risen continuously, as one can remarkably improve the minimal processed feature size and throughput with different laser parameters for different materials and different processes [1].

In the field of laser micromachining, the concept of ablation-cooling was introduced by Fatih Ömer Ilday's team [2,3]. Using femtosecond pulses and very high pulse repetition rates in GHz range, which enables the ablation-cooling process, it was demonstrated micromachining with an extremely low heat-affected zone. This principle reduces the single pulse energy required for the process and has a broad range of possible scientific, medical and industrial applications.

Single and multimode diode pumped solid state ultrafast lasers in the GHz pulse repetition rate regime have been demonstrated previously, but each of them with very different laser cavity and not tunable in repetition rate [4–7]. In our work, we demonstrate a novel concept for a fundamentally modelocked laser cavity, which allows to continuously tune the pulse repetition rate from 0.5 to 1.3 GHz with the same optical components on a simple mechanical platform. Furthermore, we demonstrate that our femtosecond laser has a very low noise level, considering that the laser was an open breadboard setup.

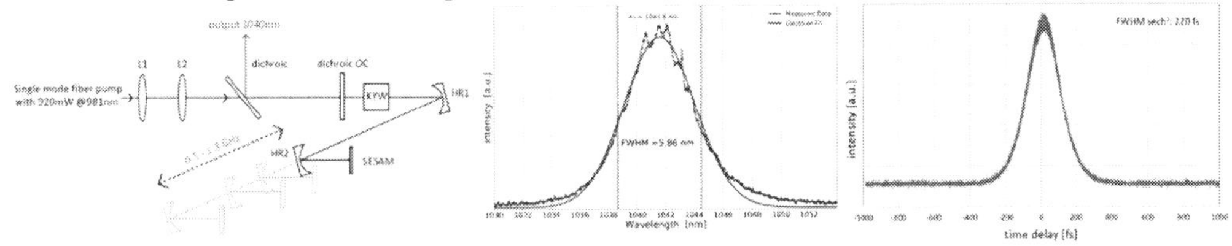

Fig. 1 Laser cavity setup (left), optical spectrum (middle) and SHG intensity autocorrelation trace (right) of a 1.23 GHz pulse repetition rate cavity

The Yb:KYW crystal based cavity emits 230 femtosecond pulses at a central wavelength around 1040 nm, with a self-starting SESAM modelocking. In a robust modelocked operation mode, an average power up to 200 mW is achieved using a stabilized single mode pump diode at a wavelength of 981 nm, emitting 920 mW average power, while the laser in continuous wave (CW) mode delivers up to 400 mW with the pump power of 920 mw. Atmospheric air cooling was sufficient for both CW and modelocked laser operation.

The developed laser will be used for investigation of the performance of 100 W-level amplifiers [8] and GHz fast modulators to achieve 100 W, GHz repetition rate, fs laser with single pulse switching capability. Moreover, this laser will be used as a widely tunable pulse repetition rate seed source for in-depth studies of micro machining in the regime of ablation-cooling, and other application where continuous tunability of the pulse repetition rate is beneficial. Furthermore, the low noise laser seems promising as a clocking signal for synchronization of other signal sources in various applications.

References

[1] R. Stoian and J.-P. Colombier, "Advances in ultrafast laser structuring of materials at the nanoscale," Nanophotonics **9**, 4665–4688 (2020).

[2] H. Kalaycioglu, P. Elahi, O. Akcaalan, and F. Ilday, "High-repetition-rate ultrafast fiber lasers for material processing (Invited paper)," IEEE Journal of Selected Topics in Quantum Electronics **24**, (2018).

[3] C. Kerse, H. Kalaycıoğlu, P. Elahi, B. Çetin, D. K. Kesim, Ö. Akçaalan, S. Yavaş, M. D. Aşık, B. Öktem, H. Hoogland, R. Holzwarth, and F. Ö. Ilday, "Ablation-cooled material removal with ultrafast bursts of pulses," Nature **537**, 84–88 (2016).

[4] P. Wasylczyk, P. Wnuk, and C. Radzewicz, "Passively modelocked, diode-pumped Yb:KYW femtosecond oscillator with 1 GHz repetition rate," Opt. Express, OE **17**, 5630–5636 (2009).

[5] A. Agnesi, A. Greborio, F. Pirzio, G. Reali, J. A. der Au, and A. Guandalini, "40-fs Yb^{3+}:$CaGdAlO_4$ laser pumped by a single-mode 350-mW laser diode," Opt. Express, OE **20**, 10077–10082 (2012).

[6] S. Pekarek, A. Klenner, T. Südmeyer, C. Fiebig, K. Paschke, G. Erbert, and U. Keller, "Femtosecond diode-pumped solid-state laser with a repetition rate of 4.8 GHz," Opt. Express, OE **20**, 4248–4253 (2012).

[7] Z. Y. Zhang, A. E. H. Oehler, B. Resan, S. Kurmulis, K. J. Zhou, Q. Wang, M. Mangold, T. Süedmeyer, U. Keller, K. J. Weingarten, and R. A. Hogg, "1.55 µm InAs/GaAs Quantum Dots and High Repetition Rate Quantum Dot SESAM Mode-locked Laser," Sci Rep **2**, 477 (2012).

[8] V. Markovic, A. Rohrbacher, P. Hofmann, W. Pallmann, S. Pierrot, and B. Resan, "160 W 800 fs Yb:YAG single crystal fiber amplifier without CPA," Opt Express **23**, 25883–25888 (2015).

 https://doi.org/10.1051/epjconf/202226701006

Highly efficient cavity-dumped Q-switched Alexandrite laser

Stefanie Unland[1], Roland Kalms[1], Peter Weßels[1], Dietmar Kracht[1] and Jörg Neumann[1]
[1]*Laser Zentrum Hannover e.V., Hollerithallee 8, 30419 Hannover, Germany*

Alexandrite as laser active medium has many benefits for spaceborne earth observation missions, e.g. a tunable laser wavelength in the range of 700 to 850 nm, a high thermal conductivity (23 $Wm^{-1}K^{-1}$) and a good breaking strength [1]. Another advantage is the high efficiency due to the small quantum defect, especially if used in laser diode-pumped systems. Therefore, Alexandrite lasers are good candidates to be used in LIDAR applications like altimetry or vegetation monitoring, in which pulse durations in the nanosecond range and high pulse repetition frequencies (PRF) are required.

We present an enhanced cavity-dumped Q-switched Alexandrite laser with CW double-pass diode pumping to achieve high-efficiency and short-pulse laser operation in a PRF range of 1-20 kHz. For the L-shaped laser system (see Fig. 1a), a 0.22 at%-doped Alexandrite crystal with an aperture of 4 x 4 mm^2 and a length of 10 mm was embedded in a thermoelectric cooler controlled copper mount to tune the crystal temperature between 25 °C and 120 °C. The crystal was pumped by a fiber-coupled laser diode offering maximum 43 W at a center wavelength of 638 nm. A retro-reflecting pump-optic (PO) was used to optimize the amount of pump absorption (90 %). The beam at the focus position inside the crystal had a super-Gaussian shape with a diameter of 320 µm. The cavity was implemented using an input coupler (IC), a bending mirror (BM) and a highly reflecting end mirror (HR). Cavity-dumped Q-switching was achieved by an integrated BBO Pockels cell in combination with a quarter-wave plate (QWP) and a thin-film polarizer (TFP). The laser pulse inside the cavity built up during the switching-on time of the Pockels cell and was coupled out at the TFP by rapidly switching off the high voltage. During the measurements, the delay between switching on and off as well as the axial pump focus position was optimized to reach the maximum output energy.

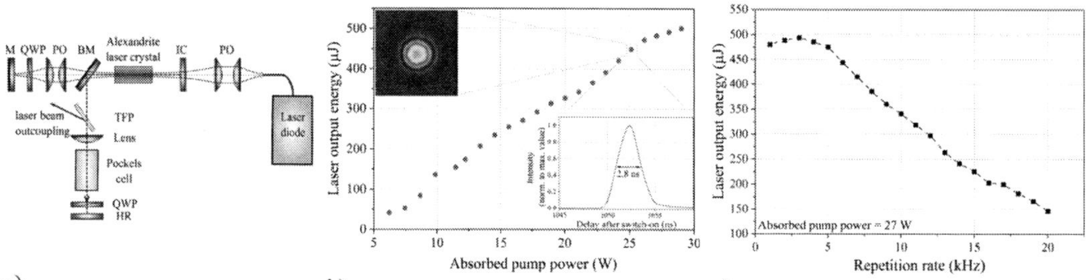

a) b) c)

Fig. 1 a) Schematic drawing of the cavity-dumped Q-switched Alexandrite laser with double-pass diode pumping; b) Laser output energy as a function of the absorbed pump power, inset: temporal pulse shape with 2.8 ns FWHM and beam profile at around 25 W absorbed pump power; c) Variation of the pulse energy with PRF, measured for an absorbed pump power of 27 W.

The laser output characteristics achieved with the setup are shown in Fig. 1b for a PRF of 5 kHz. A maximum output energy of more than 500 µJ (average power of 2.5 W) at 755 nm with a pulse duration of 2.8 ns could be achieved, which results in a record peak power exceeding 179 kW reached within a cavity-dumped Q-switched Alexandrite laser with CW diode pumping (more than twice of the value found in literature [2]). Moreover, the optical-to-optical efficiency could be improved from 5.0 % to 8.2 % using the fiber-coupled laser diode with a super-Gaussian beam profile at the pump focus position inside the crystal. The beam profile of the laser output, shown in the inset of Fig. 1b, revealed a TEM_{00} mode beam with $M_x^2 = 1.33$ and $M_y^2 = 1.42$. During a long-term stability measurement at an output energy of around 350 µJ with 5 kHz PRF, no degradation over 150 megashots was observed. The variation of the pulse energy with the PRF adjusted in the range of 1 kHz to 20 kHz at a fixed pump power of 27 W is illustrated in Fig. 1c. The resulting curve shows a maximum at 3 kHz and a nearly linear decrease afterwards. This is the first time that cavity-dumped Q-switched Alexandrite laser operation was demonstrated for PRF over 10 kHz.

The presented work was carried out in the course of the GALACTIC project, which has received funding from the European Union's Horizon 2020 research and innovation programme under grant agreement No 870427 [3].

References

[1] J. C. Walling, O. G. Peterson, H. P. Jenssen, R. C. Morris, and E. W. O'Dell, "Tunable Alexandrite Lasers," IEEE J. Quant. Electron. 16 (12), 1302–1315 (1980).
[2] G. M. Thomas, A. Minassian, X. Sheng, and M. J. Damzen, "Diode-pumped Alexandrite lasers in Q-switched and cavity-dumped Q-switched operation", Optics Express 24 (24), 27212 (2016).
[3] Cordis Europa, Horizon2020, Fact Sheet: "High Performance Alexandrite Crystals and Coatings for High Power Space Applications.", https://cordis.europa.eu/project/id/870427

 https://doi.org/10.1051/epjconf/202226701007

Pr:$YAlO_3$ microchip lasers operating at crystal temperatures close to liquid helium temperature

M.Fibrich[1,2]***, J. Šulc***[1]***, H. Jelínková***[1]

1. Czech Technical University in Prague, Faculty of Nuclear Sciences and Physical Engineering Břehová 7, 115 19 Prague 1, Czech Republic
2. Institute of Physics ASCR, ELI-Beamlines, Na Slovance 1999/2, 182 21 Prague, Czech Republic

In this contribution, Pr:YAP laser performance enhancement by operating the laser close to liquid helium temperature is reported; watt-levels of the laser output in the near-infrared, orange, green, and cyan-blue spectral region are presented under the 4-W InGaN LD pumping.

To realize Pr:YAP laser systems, the microchip resonator geometry formed by a cavity mirrors directly deposited on the crystal faces was proposed. This geometry allows to minimize resonator losses and it is also very attractive due to its compact and rugged design. Four microchip crystals of the same active ion concentrations (c = 0.6 at. %), dimensions (ϕ5× 5 mm), cuts (along the b-axis according to Pbnm notation), and the output facet reflectivities of R = 98% at the respective wavelengths of interest (around 747 nm, 622 nm, 547 nm, and 493 nm) were used. The microchip pump sides (pump mirror) were highly transmissive for the LD pump radiation, and highly reflective (R > 99.8%) for the generated radiations. Besides, special attention was paid to a suppression of the laser oscillations at the orange laser transition (except the microchip crystal designed for the orange emission) because the emission cross section in the orange (622 nm) is e. g. two times higher than that in the red (747 nm), and ten times higher than that in the green (547 nm) at the very low crystal temperatures. All Pr:YAP crystals were grown by the Czochralski method. The microchip crystal was mounted on the cold-finger of the He-cryostat (OptistatDry BLV cryostat; Oxford Instruments) with uncoated CaF2 windows. The crystal temperature could be set at any value within the 4 – 300 K range. Blue InGaN LD (Nichia Corporation) with the maximum available output power of 4 W at 447 nm wavelength in a linearly polarized beam was used as a pump source. The pump LD radiation collimated by an aspherical lens (f = 4.5 mm, 0.55 NA, ARC = 400–600 nm) was focused into the microchip crystal by a lens with the 60 mm focal length, resulting in a spot size of about 60 μm in radius. A half-wave plate designed for the 400–800 nm wavelengths (Thorlabs Inc.) was employed to set the LD beam polarization parallel with the "c"-crystallographic axis of the Pr:YAP crystal, which provides the higher absorption efficiency if compared with E||a crystal orientation. Laser output radiation was detected behind the cut-off filter.

The Pr:YAP input-output characteristics at 5 K crystal temperature for the 747 nm, 622 nm, 547 nm, and 493 nm emission are displayed in Fig. 1. The highest output power amounting to 1.5 W was reached at 747 nm laser transition. The corresponding oscillation threshold and slope efficiency with respect to the absorbed pump power were 300 mW and 46%, respectively. The Pr:YAP lasers behaviors in terms of the maximal output power as a function of the crystal temperature within the 5–300 K range are seen in Fig. 2.

It is obvious, the crystal temperature has a strong impact on the Pr:YAP output properties. The laser performance enhancement is driven not only by the increase of the emission cross section while the crystal temperature is going down, but also by the involved temperature dependent energy transfer processes, such as cross-relaxation (orange laser transition) or reabsorption processes (blue laser transition).

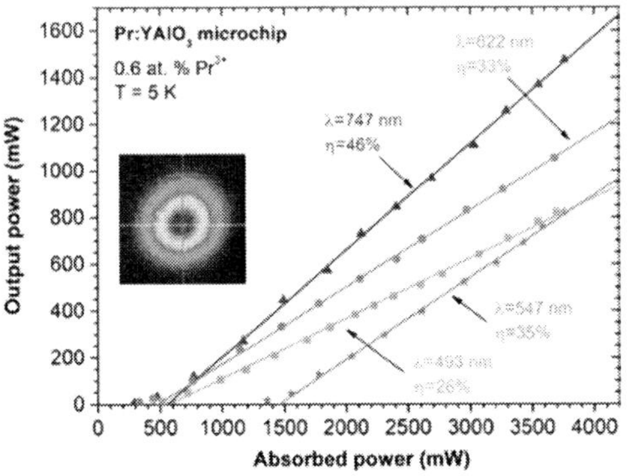

Fig. 1. Output characteristics of Pr:YAP microchip lasers designed for generation in near-infrared, orange, green, and cyan-blue laser transitions at 5 K crystal temperature; inset – spatial beam profile at maximal output power for illustrative purposes.

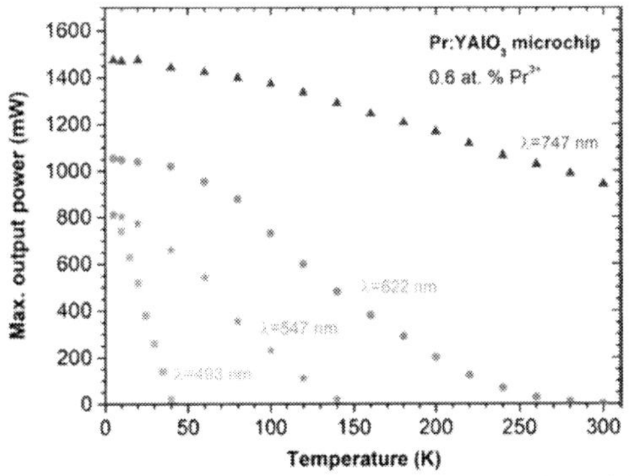

Fig. 2. Maximal output powers of Pr:YAP microchip lasers designed for generation in near-infrared, orange, green, and cyan-blue laser transitions as a function of crystal temperature.

Terahertz radiation in tailored two-color laser fields with a stabilized doubly resonant optical parametric oscillator

H. Rao[1,2], C. M. Dietrich[1,2], J. R. C. Andrade[3], A. Demircan[1,2], I. Babushkin[1,2,3], and U. Morgner[1,2]
[1]*Leibniz University Hannover, Institute of Quantum Optics, Welfengarten 1, 30167 Hannover, Germany*
[2]*Cluster of Excellence PhoenixD, Welfengarten 1, 30167 Hannover, Germany*
[3]*Max Born Institute, Max Born Str.2a, 12489 Berlin, Germany*

Strong tailored two- and three-color optical waveshapes can be useful for effective generation of light at very high (XUV) and very low (THz) frequencies [1,2]. In particular, for generation of THz radiation, strong and stable asymmetric temporal-waveshapes are needed. Phase locked doubly-resonant optical parametric oscillators (DROPOs) can contribute with intensities high enough for this goal. In this work, we stabilize our degenerate DROPO by using a locking scheme which utilizes monitoring of a "spurious"-sum-frequency product (SFG) of signal and pump in the amplifier crystal – a method proposed very recently in [3]. In our experiment, we synchronously pump the DROPO by a home-built Yb:YAG Kerr-lens mode-locked thin-disk laser, emitting pulses at a wavelength of 1030 nm with a pulse duration of 250 fs, 20 W output power and 33 MHz repetition rate.

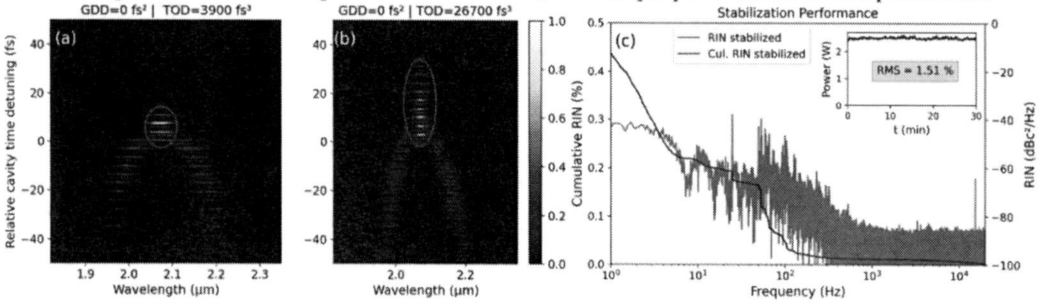

Fig. 1. Extending the self-locking region by changing the third-order dispersion (TOD). (a), (b) Experimental cavity length scans for small and large values of TOD, respectively. The self-locking region is marked with the red ellipse. (c) Relative intensity noise (RIN) of the signal output of the DROPO for active PI controller (red). The cumulative RIN for the stabilized system is less than 0.9% (blue). Inset shows longtime measurement of the output power.

For the optimum performance of the stabilization, a large self-locked region of the spectrum is beneficial. Here we show that the phase locking region can be significantly extended by introducing high-order dispersion (TOD) into the cavity [4]. As shown in Fig.1 (a) and (b), the self-locked region is highly dependent on TOD: with increasing TOD, the group-velocity matching parabola becomes narrower, the self-locked region increases.

Furthermore, we use SFG as the error signal for the stabilization of the system which can be modified by spectral filtering for different incident angles. From Fig. 1(c) we can see that this approach allows to reduce the noise drastically for all frequencies below -40 dBc^2/Hz. The cumulative RIN is here below 0.5% which is quite low. The output power is stabilized for an extended time periods of order of 30 minutes and more.

In conclusion, the introduction of TOD in the cavity can significantly increase the self-locking region of DROPO. Also, we were able to stabilize our DROPO using an SFG-based scheme. With all this, as we believe, we made an important step towards creating steady multi-color waveforms for intracavity strong-field optics.

References

[1] I. Babushkin, C. Bree, C. M. Dietrich, A. Demircan, U. Morgner and A. Husakou, "Terahertz and higher-order Brunel harmonics: from tunnel to multiphoton ionization regime in tailored fields," J. Mod. Opt., 64, 1078-1087 (2017).

[2] D. Jang, R. M. Schwartz, D. Woodbury, J. G. M. Mahon, A. H. Younis, H. M. Milchberg and K. Y. Kim, "Efficient terahertz and Brunel harmonic generation from air plasma via mid-infrared coherent control," Opt., 6, 1338-1341 (2019).

[3] Y. S. Cheng, R. A. McCracken, and D. T. Reid, "Dither-free stabilization of a femtosecond doubly resonant opo using parasitic sum-frequency mixing," Opt. Lett., 45, 768-771, (2020).

[4] C. M. Dietrich, I. Babushkin, J. R. C. Andrade, H. Rao, A. Demircan and U. Morgner, "Higher-order dispersion and the spectral behavior in a doubly resonant optical parametric oscillator", Opt. Lett., 45, 5644-5647 (2020).

Spatially-multiplexed tunable dual-comb optical parametric oscillator at 250 MHz

C. P. Bauer,* J. Pupeikis, B. Willenberg, Z. A. Bejm, N. Pezzoli, C. R. Phillips and U. Keller
Institute of Quantum Electronics, Eidgenössische Technische Hochschule Zürich, Auguste-Piccard Hof 1, 8093 Zurich, Switzerland
**cabauer@phys.ethz.ch*

Abstract: We demonstrate a spatially-multiplexed dual-comb 250-MHz OPO from a single linear cavity. The adjustable repetition-rate difference is 4.1 kHz. Each idler comb has >200-mW average power at 3.5 μm with 30 nm bandwidth. The OPO is wavelength-tunable from 1.36 μm to 1.7 μm and 2.9 μm to 4.17 μm. © 2022 The Author(s)

Dual optical frequency combs in the mid-IR are valuable tools for spectroscopic measurements in the so-called molecular fingerprint region [1]. Among the many technologies for mid-IR generation, optical parametric oscillators (OPOs) pumped by solid-state lasers (SSLs) are compelling due to their flexible pulse parameters and a wide center wavelength tuning range enabled by the parametric process. Additionally, spectroscopic measurement signal-to-noise ratio is sensitive to intensity noise, where ultrafast OPOs are advantageous [2]. Yet, most dual-comb OPOs are complex systems including at least three different cavities [3]. Here, we demonstrate a single-cavity dual-comb OPO synchronously and collinearly pumped by a single-cavity dual-comb Yb:YAG laser with a similar architecture to the one presented in [4]. While the design of our previous bidirectional OPO limited us to 125 Hz repetition rate difference Δf_{rep} and a fixed wavelength at a low repetition rate of f_{rep}=80 MHz [2], the new OPO presented here is spatially multiplexed in a linear cavity with the help of a new intra-cavity biprism scheme [4]. Moreover, we use picosecond pulses at a repetition rate of f_{rep}=250 MHz, which supports high-spectral resolution together with a high repetition rate difference of 4.1 kHz. This enables a simple, broadly tunable, and high-power-per-comb-line source that can serve as a tool for dual-comb spectroscopy at high signal-to-noise ratio.

The schematic of our dual-comb OPO cavity is illustrated in Fig. 1(a). Both the pump and the OPO are designed such that the two combs take a similar path through the cavity sharing all the optical elements, and thus most noise sources added by the cavity are expected to be correlated.

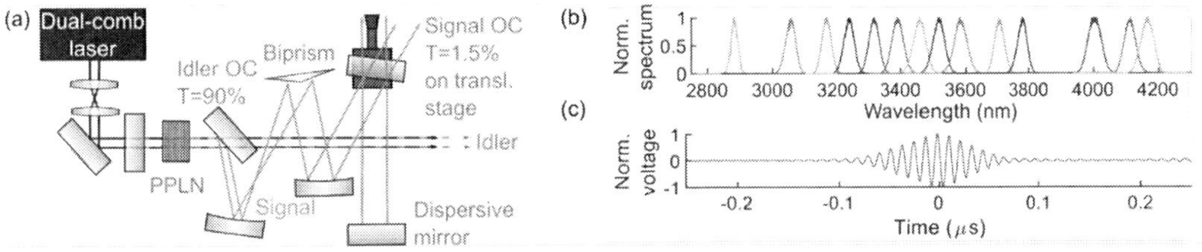

Fig. 1. (a) Schematic of the linear OPO cavity. Note, that for clarity the two combs are shown horizontally separated. In the laboratory, they are vertically multiplexed. (b) Tunable optical spectra of the idler combs measured with a Thorlabs OSA205. (c) Typical interferogram of the two signal beams in the time domain.

Up to 1.6 W of average power per comb at a center wavelength of 1031 nm, and a FWHM optical bandwidth of 1.7 nm (pump 1) and 1.8 nm (pump 2) is directed to the common pump focusing lens. The OPO uses a periodically poled lithium niobate (PPLN) crystal which is 4-mm-long, 5 mol. % MgO doped, and has a fan-out grating with periods ranging from 24.8 μm to 31.2 μm (HC Photonics). The residual pump and idler combs are transmitted via the idler output coupler (OC). The biprism introduces a cavity length difference for the two signal combs such that the OPO can run at high and tunable repetition rate differences. The synchronicity of the signal cavity with the pump is obtained by scanning the OPO cavity length with a translation stage. Dispersive mirrors are used to reduce the OPO sensitivity to cavity length fluctuations.

The OPO delivers idler average powers of 200 mW (idler 1) and 220 mW (idler 2), and total combined signal powers of 360 mW (signal 1) and 370 mW (signal 2) at an idler wavelength of 2.9 μm and signal wavelength of 1.6 μm after their respective OC. Due to the relatively narrow bandwidth and high repetition rate, the OPO idler has a power per comb line of about 60 μW. The broad tuning range of the idler from 2.9 μm to 4.17 μm is shown in Figure 1(b). Figure 1(c) shows a typical interferogram from the beating of the two signal combs uppon combination at a non-polarizing beam splitter, illustrating the dual-comb nature of the source.

5. References:

1. I. Coddington, N. Newbury, and W. Swann, "Dual-comb spectroscopy," Optica **3**, 414-426 (2016).
2. C. P. Bauer, S. L. Camenzind, J. Pupeikis, B. Willenberg, C. R. Phillips, and U. Keller, "Dual-comb optical parametric oscillator in the mid-infrared based on a single free-running cavity," Opt. Express **30**, 19904-19921 (2022).
3. Z. Zhang, C. Gu, J. Sun, C. Wang, T. Gardiner, and D. T. Reid, "Asynchronous midinfrared ultrafast optical parametric oscillator for dual-comb spectroscopy," Opt. Lett. **37**, 187-189 (2012).
4. J. Pupeikis, B. Willenberg, S. L. Camenzind, A. Benayad, P. Camy, C. R. Phillips, and U. Keller, "Spatially multiplexed single-cavity dual-comb laser for equivalent time sampling applications," arXiv, 2203.08536 (2022).

Influence of Disk Aberrations on High-Power Thin-Disk Laser Cavities

Moritz Seidel , Lukas Lang, Christopher R. Phillips and Ursula Keller
Department of Physics, Institute for Quantum Electronics, ETH Zurich, 8093 Zurich, Switzerland

Introduction: High-power thin-disk lasers (TDLs) are commonly used in industrial, medical, and academic contexts both in continuous-wave (cw) or modelocked operation. Enabled by the thin-disk gain geometry, the power-scaling of TDLs is accomplished by simultaneously increasing the pump and laser mode size. However, as the sensitivity to thermal effects scales with the laser mode size on the disk [1], it is critical to understand these effects in order to further scale the performance of high-power ultrafast TDL oscillators [2].

Thin-disk oscillator cavities are often modeled using ray transfer matrix analysis by computing their ABCD matrix and solving for the complex beam parameter. This analysis can include temperature-dependent radii of curvature, but always assumes parabolic surface profiles. The reduction in beam quality due to aberrations in the cavity can thus not be captured. Here we present a numerical, spatially resolved split-step model that supports arbitrary non-parabolic surface profiles for cavity elements such as the thin-disk. We find that these aberrations have a critical impact on the resulting cavity mode shape. Our results show good qualitative agreement between experiment and simulations and are an important step to scaling modelocked thin-disk lasers to the kW-level.

Model and setup: Our model finds an optical resonator's steady-state electric field distribution by simulating many cavity roundtrips. The cavity is described as a series of cavity elements separated by free space propagation. For each cavity element, spatially varying amplitude and phase factors are applied to the electric field. We approximate the spherical mirrors with a parabolic phase profile. For the thin-disk we use phase profiles measured during laser operation with a Twymann-Green interferometer (TRIOPTICS μPhase-500). The resulting power-dependent surface profile is decomposed into Zernike polynomials up to the $n + |m| = 6$-th order and supplied to the simulation (illustrated in Fig. 1(a) with parabolic parts removed). While the disk is almost parabolic, some non-radially-symmetric aberrations are clearly visible towards the edge of the disk. Commercial kW-class lasers employ advanced proprietary thin-disk technology with even better flatness and close to perfect radial symmetry, which have enabled near-fundamental mode output powers beyond 10 kW in cw-operation [3], and 2.5 kW in ultrafast multipass amplifiers [4]. The amplitude factor of the disk is computed by solving the quasi-three-level rate equations for the laser gain medium including the pump and laser intensity.

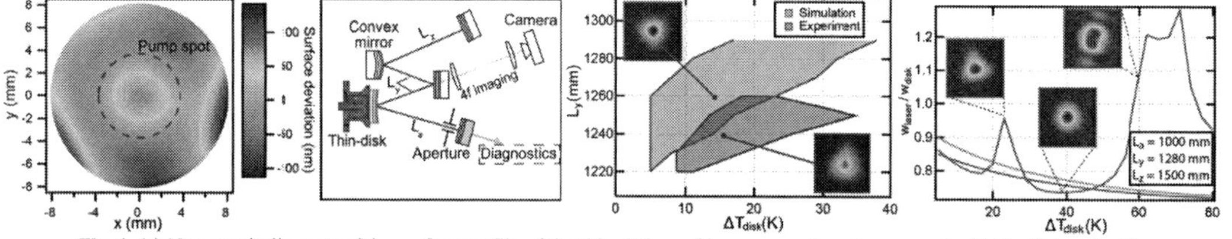

Fig. 1. (a) Non-parabolic parts of the surface profile of the thin-disk used in our experiment measured with the disk optically pumped to 73.9 °C. (b) the cavity configuration which includes a hard aperture. The arm lengths L_a, L_y and L_z are indicated. The laser mode on the disk is imaged onto a camera. (c) Comparison of simulation and experiment for cavities that produce triangular distorted fundamental modes with L_z=1500 mm. Insets show measured and simulated beam profiles. (d) Simulation of a cavity with arm length as indicated comparing the ray matrix approach (green), and our model for an ideal parabolic thin-disk (orange) and the disk with the measured phase profile (blue).

Results: The model is validated by experiment in a convex-concave thin-disk laser cavity (Fig. 1(b)). The convex mirror has a radius of curvature of -2 m. The thin-disk used is an Yb-doped disk contacted by TRUMPF Scientific Lasers, with a cold radius of curvature of 3.84 m. An output coupler with 10% transmission is used. The simulations predict distorted fundamental modes for certain cavity configurations, which were subsequently experimentally tested. Fig. 1(c) shows scans of T_{disk} (via changes to the pump power) and L_y (distance disk – convex mirror) for both the simulation and experiment. In the highlighted areas, we observe somewhat triangularly distorted fundamental modes, likely stemming from the trefoil symmetry of the disk phase profile (Fig. 1(a)). While the overall trends are well-reproduced, we observe a shift of the predicted zone to shorter L_y when comparing to simulation, potentially due to offsets in the curvature of the cavity elements. Fig. 1(d) shows the drastically different behavior predicted by our model when compared to simulations assuming a parabolic phase profile.

Conclusion: We present a spatially-resolved numerical model backed by experimental data to predict the output mode of a thin-disk laser oscillator. By including the disk's measured phase with its non-parabolic aberrations, we recover experimentally observed beam quality degradations not captured by simpler models.

References:

[1] V. Magni, "Multielement stable resonators containing a variable lens," JOSA A **4**, 1962-1969 (1987).
[2] F. Saltarelli et al. , "Power scaling of ultrafast oscillators: 350-W average-power sub-picosecond thin-disk laser," Opt Express **27**, 31465-31474 (2019).
[3] S. Nagel et al. , "Thin-disk laser system operating above 10 kW at near fundamental mode beam quality," Opt. Lett. **46**, 965-968 (2021).
[4] J. Dominik et al. , "Multi-kilowatt Ultrafast Laser with Thin-disk Technology," in *Laser Congress 2021 (ASSL,LAC)* AM2A.6.

Cryogenically cooled compact Yb:Lu_2O_3 laser

Venkatesan Jambunathan,[1,*] Bruno J Le Garrec,[2] Martin Smrz,[1] and Tomas Mocek[1]
[1]*HiLASE Center, Institute of Physics of the Czech Academy of Sciences, Za Radnicí 828, 25241 Dolní Břežany, Czech Republic*
[2]*Laboratoire pour l'Utilisation des Lasers Intenses (LULI), Unité Mixte n° 7605 CNRS - CEA - Ecole Polytechnique - UPMC, Route de Saclay, 91128 Palaiseau Cedex, France*
**Corresponding author: jambunath@fzu.cz*

In recent years, compact diode pumped solid-state lasers based on Ytterbium (Yb^{3+})-doped ions have attracted the scientific community due to their vast scientific and technological applications. Nevertheless, at room temperature, the Yb^{3+} doped gain materials behave as quasi three-level lasers, due to the finite population at ground state leading to reabsorption at laser wavelength and mitigate the power scaling capabilities. To overcome this issue and to enhance the output, the gain medium has to be cooled down to cryogenic temperatures, where four level is accomplished leading to an overall laser performance [1] . Many host materials have been studied in the past, here in this work we considered Lu_2O_3, which belongs to the cubic sesquioxides family. This host is known for its excellent thermal properties and it make it attractive for high power solid-state lasers [2].

The laser experiments were carried out in a compact modular cryogenic vacuum chamber that has the provision to mount the optics and sample as close as possible to form a compact laser cavity. As gain medium, an un-coated 5×5 mm^2 aperture and 3 mm thickness Yb: Lu_2O_3 ceramics of 1 at.% was used. It was mounted in a copper holder and was placed in between two mirrors to form a compact plano- plano cavity. For cooling down to cryogenic temperature, a two stage closed cycle cryostat having a heat load of 13 W at 100 K and to maintain the sample temperature, a 50 Ω heater connected near the cold finger were used. A VBG stabilized fiber coupled diode laser emitting at 976 nm with a spectral bandwidth of 0.40 nm (N.A. = 0.22, core diameter: 105 μm) was used as a pump source. The un-polarized pump beam was imaged using achromatic lenses with a ratio of 1:1.5.

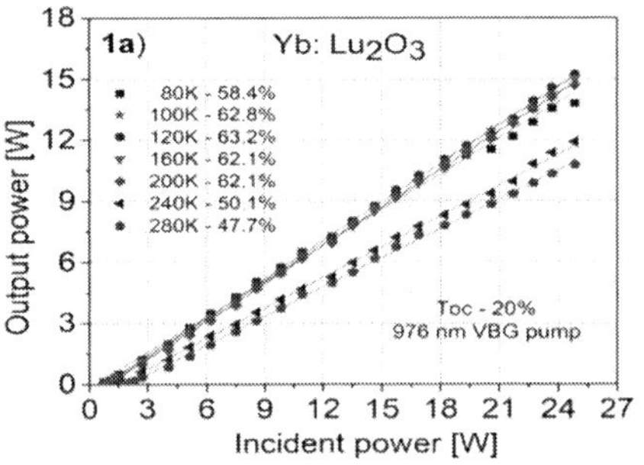

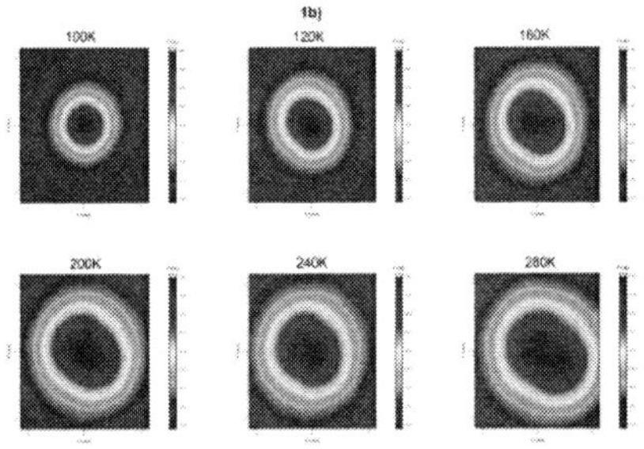

Fig. 1 (a) Input - Output power characteristics of 1. at.% Yb:Lu_2O_3 ceramic at different temperatures,T_{OC} = 20%; (b) Obtained beam profile for different temperatures at maximum power.

Continuous wave (CW) laser operation was realized for all the temperatures with 20% Toc in step size of 20 K from 80 K until 280 K. In all cases, the output power increases and the laser threshold decreases with decrease in temperature as shown in Fig. 1 (a). A maximum output power of 15.23 W for an incident power of 24.85 W was acheived for 120 K corresponding to a slope efficiency of 63% and the laser emits at 1031.6 nm. To be noted, from the results, although the output power and slope efficiency from 80 – 200 K were close as seen in Fig 1(a), but the beam profile changes from single mode to multi-mode (see Fig 1 (b)). From this, one can conclude that the cryogenic cooling has a huge impact on not only achieving better output but also on achieving a very good beam profile. Present work is focused on using other transmission of output coupling.

References

[1] T. Y. Fan, D. J. Ripin, R. Aggarwal, J. R. Ochoa, B. Chann, M. Tilleman, and J. Spitzberg, "Cryogenic Yb3+-Doped Solid-State Lasers", IEEE J. Sel. Top. Quantum Electron. **13**, 448-459 (2007).
[2] D. Rand, D. Miller, D. J. Ripin, and T. Y. Fan, "Cryogenic Yb^{3+}-doped materials for pulsed solid-state laser applications," Opt. Mater. Express 1(3), 434–450 (2011).

Thermo-optical wavefront distortions in Nd:YVO_4 laser amplifiers

Merle Schneewind[1,3], Phillip Booker[1,3], Sergii Iakushev[1,3], Peter Weßels[1,3], Benno Willke[2,3], Jörg Neumann[1,3], and Dietmar Kracht[1,3]

1. Laser Zentrum Hannover e.V., Hollerithallee 8, 30419 Hanover, Germany
2. Max-Planck-Institut für Gravitationsphysik (Albert-Einstein-Institut), Callinstr. 38, 30167 Hanover, Germany
3. Cluster of Excellence QuantumFrontiers, Welfengarten 1, 30167 Hanover, Germany
m.schneewind@lzh.de

Abstract: The power dependence of a Nd:YVO_4 laser amplifier beam wavefront was analyzed by Zernike polynomial decomposition. This analysis was performed experimentally and by simulations based on split-step Fourier propagation showing a good agreement. The simulations yield a base for the design of an aberration compensation system.

Nd:YVO_4 laser crystals are promising laser technology for future gravitational wave detectors (GWDs) [1] because of several reasons. Nd:YVO_4 features anisotropic emission cross-sections and inherent birefringence enabling stable output polarization and suppressing thermally induced depolarization. Furthermore, Nd:YVO_4 features a low saturation intensity. A major obstacle for the application of Nd:YVO_4 laser crystals, however, is vanadate's lower thermal conductivity and its higher brittleness compared to YAG [2]. Thus, detrimental thermal effects through thermo-optical coupling and the associated stress can exacerbate the power scaling process of Nd:YVO_4 laser crystals.

In this work, we established a numerical simulation tool to predict beam parameters of Nd:YVO_4 amplifiers including power slope, efficiency and the magnitude of thermal wavefront degradations to design a suitable aberration compensation system. The simulation tool is based on iterative computation of the complex field propagation via a split-step Fourier beam propagation model and a finite-difference thermal equation solver. The resulting wavefront was analysed by Zernike polynomial decomposition. The simulation input parameters are matched to an experimental amplifier with an a-cut Nd:YVO_4 crystal (3 mm x 3 mm x 12 mm, 0.15 at.%) which was seeded with 1.6 W single-frequency light at 1064 nm wavelength.

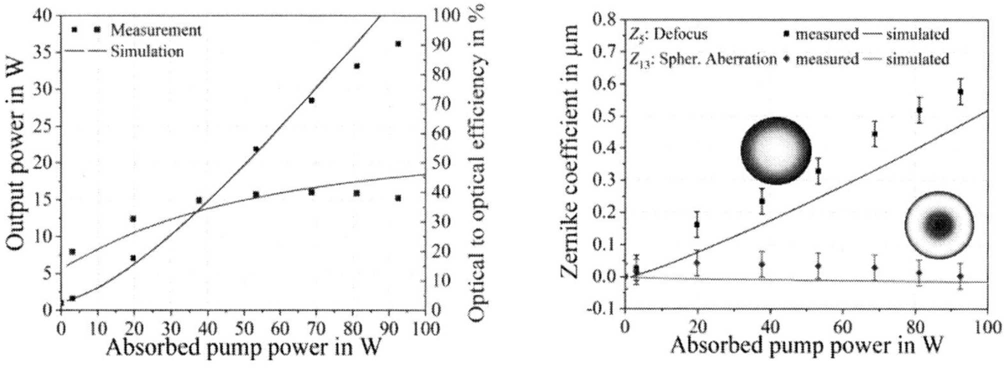

Fig. 1, left: power slope and optical-to-optical efficiency, right: symmetrical Zernike coefficients versus absorbed pump power

The simulated and measured output powers and efficiencies shown in Fig. 1, left are in good agreement. We obtained output powers up to more than 30 W (69 W absorbed pump power) in a single pass with around 45 % efficiency. The first symmetrical Zernike coefficients are depicted in Fig. 1, right; the corresponding Zernike polynomials are shown as the inset. The experimental and simulated values of Z_5, i.e. defocus, are in excellent agreement fortifying the validity of the simulated results. Furthermore, the next symmetrical term Z_{13}, i.e. spherical aberration, is close to zero in both experiment and simulation.

1. N. Bode, F. Meylahn, and B. Willke, "Sequential high-power laser amplifiers for gravitational wave detection", Opt. Express **28** (2020)
2. I. Zawischa, M. Brendel, K. Danzmann, C. Fallnich, M. Heurs, S. Nagano, V. Quetschke, H. Welling, and B. Willke, „The GEO 600 laser system", Class. Quantum Grav. **19** (2002)

EPJ Web of Conferences **267**, 01013 (2022)
EUROPHOTON 2022

https://doi.org/10.1051/epjconf/202226701013

Comparative study on pump wavelength dependent efficiency in Nd:YVO_4

Merle Schneewind[1,2], Stefan Spiekermann[1], Peter Weßels[1,2], Jörg Neumann[1,2], and Dietmar Kracht[1,2]

1. Laser Zentrum Hannover e.V., Hollerithallee 8, 30419 Hanover, Germany
2. Cluster of Excellence QuantumFrontiers, Welfengarten 1, 30167 Hanover, Germany
m.schneewind@lzh.de

Abstract: The influence of the pump wavelength on the heat load and efficiency of Nd:YVO_4 crystals is investigated with a specially designed crystal mount. The measurements indicate that the change in heat load in the crystal can be solely ascribed to the difference of quantum defects and no further non-radiative effects.

Laser sources for future generations of gravitational wave detectors (GWDs) need to fulfil the requirements of high fundamental mode content, linear polarization and low noise properties at high power levels over 500 W at 1064 nm [1]. Nd:YVO_4 is a promising material to fulfil these requirements. Due to the large absorption cross section, Nd:YVO_4 is usually pumped at 808 nm. However, there is also an absorption line at 878 nm, which enables pumping directly into the emitting level and thereby reduces the quantum defect and the associated heat load.

In this work, we studied the efficiency of Nd:YVO_4 with a focus on the heat deposition in the crystal. For this, we developed a custom crystal mount, which is shown in Fig. 1, left. The lower part of this mount is temperature stabilized by a Peltier-element mounted on a water-cooled heat sink. The resulting heat gradient along the crystal mount is measured at different pump power levels via two temperature sensors. Before the measurement, a calibration was carried out by heating the mount via pt100 resistors. In order to achieve a similar absorption behavior along the crystal for both pump wavelengths, the 808 nm pump diode was slightly temperature shifted. This way, we achieved a similar amount of absorbed power at the same input pump power levels for both pumping schemes.

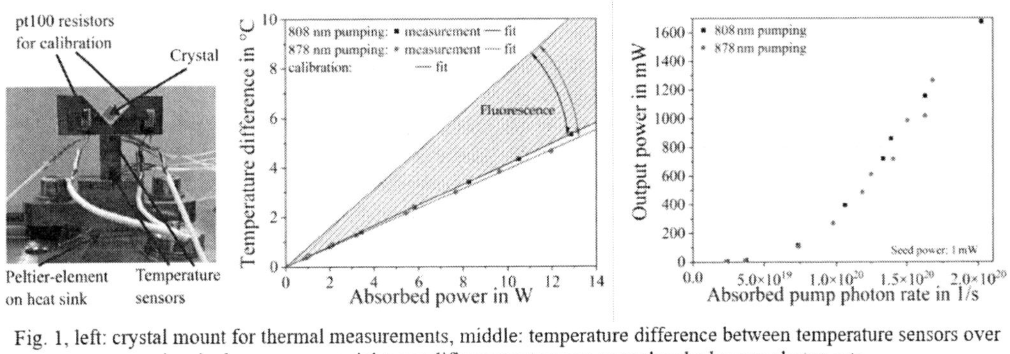

Fig. 1, left: crystal mount for thermal measurements, middle: temperature difference between temperature sensors over absorbed pump power, right: amplifier output power over absorbed pump photon rate

Fig. 1, middle shows the resulting temperature differences for different absorbed pump powers as well as the calibration curve (blue). The difference between the calibration curve and the measured values can be attributed to the fluorescence, as this is the only effect that does not contribute to the heat load. The ratio of the fluorescence measured at both pump wavelengths corrected by the quantum defect is approx. one. Thus, the difference in heat generation and in fluorescence can be explained solely by the quantum defect. Other non-radiative processes are negligible within the scope of our measurement accuracy. As this measurement was performed without any laser light extraction besides fluorescence, it was repeated with a laser resonator built around the crystal. Also here, the measured temperature difference could be explained just by the quantum defect.

Furthermore, an amplifier was set up in order to measure the pump wavelength dependent efficiency. Fig. 1, right shows the output power at different pump photon rates for both pump wavelengths. The output power per pump photon was independent from the pump wavelength within our measurement accuracy. Hence, pumping at 878 nm leads to less deposited thermal power and thus less thermal lensing effects than pumping at 808 nm at the same output power and the exact difference can be calculated by the quantum defect.

[1] ET Science Team, "Einstein gravitational wave Telescope conceptual design study" (2011), URL: https://apps.et-gw.eu/tds/ql/?c=7954

 https://doi.org/10.1051/epjconf/202226701014

Towards ultra fast pulse generation by gain-switching of diode pumped surface emitting semiconductor lasers

André Marianovich[1,*], Stefan Spiekermann[1], Moritz Brendel[2], Peter Weßels[1], Jörg Neumann[1], Markus Weyers[2], Dietmar Kracht[1]

[1]*Laser Zentrum Hannover e.V. (LZH), Laser Development Department, Hollerithallee 8, 30419 Hannover, Germany*
[2]*Ferdinand-Braun-Institut (FBH), Gustav-Kirchhoff-Str. 4, 12489 Berlin, Germany*
[*]Corresponding author: a.marianovich@lzh.de

Gain-switching is the most straightforward approach for picosecond pulse generation with semiconductor lasers. Instead of integrating any variable losses in the laser cavity, pulsed operation is achieved by modulation of the electrical or optical pump power and hence the gain. In addition to that, a general property of gain-switched semiconductor lasers is a reduction of output pulse length compared to the pump pulse due to fast pulse build-up after pumping several times above threshold [1]. Here, we will give some very first insights on our research on cascaded gain-switching of a vertical-cavity surface-emitting laser (VCSEL) optically pumped by gain-switched laser diodes. This approach will be advantageous for building a compact and robust single-frequency picosecond oscillator with kHz repetition rates.

The focus in this paper is on the development of the optically pumped VCSEL structure for picosecond pulse generation around 1030 nm. The structure design was based on the resonant periodic gain (RPG) approach [2], where double quantum well (QW) gain sections are placed at the anti-nodes of the standing wave pattern of the internal electric field. Each gain section is based on a 7 nm $In_{0.246}Ga_{0.754}As/GaAs$ double quantum well structure with strain compensation and $\lambda/2$ thickness in total.

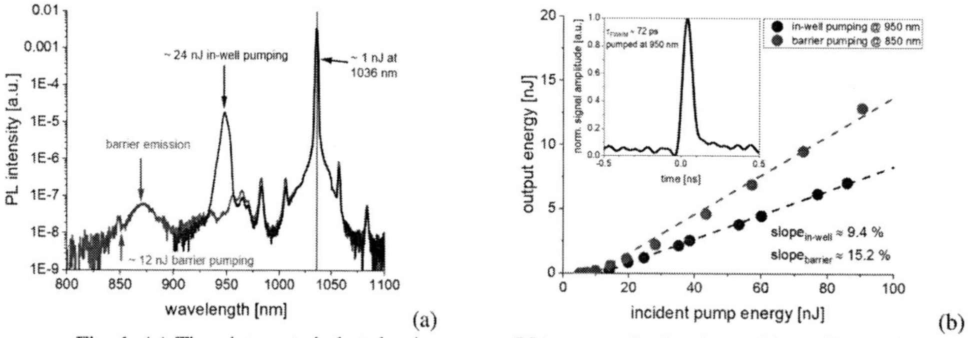

Fig. 1. (a) Time integrated photoluminescence (PL) spectra for barrier and in-well pumping of optically gain-switched VCSEL laser. (b) Slope for barrier and in-well pumping. Trace of currently shortest pulse of 72 ps FWHM (inset).

For fundamental investigations of the output parameters and their dependencies on both the pump wavelength and the pump fluence, we used a self-built wavelength tuneable pulsed laser system [3]. Excitation of discrete QW states (in-well pumping) as well as of electrons in the GaAs barriers (barrier pumping) was investigated. The goal is to replace this system with an optimized gain-switched broad area emitter in a later step. The results in Fig. 1 are from a sample with 25 stacked double-QW gain sections surrounded by a bottom DBR consisting of 25 pairs of AlAs/GaAs $\lambda/4$-layers and a top-DBR, as the output coupling mirror, consisting of 15 pairs of $AlAs/Al_{0.2}Ga_{0.8}As$ $\lambda/4$-layers with reflectivities of 99.98 % and 97.5 %, respectively, for 1030 nm. For in-well pumping at 950 nm and 86 ps pump pulse duration, output pulses at 1036 nm with energy around 1 nJ and duration of 72 ps FWHM, measured with a fast photodetector and oscilloscope, were demonstrated. Further optimization of the RPG structure and the cavity design should allow for even shorter output pulses.

References

1. P. Vasil'ev, "Ultrafast diode lasers", Artech House, Bosten, London (1995)
2. S.W. Corzine, R.S. Geels, J.W. Scott; R.-H. Yan, L.A. Coldren, "Design of Fabry-Perot surface-emitting lasers with a periodic gain structure," IEEE Journal of Quantum Electronics **25**, 1513 - 1524 (1989)
3. A. Marianovich, S. Spiekermann, M. Brendel, P. Weßels, J. Neumann, M. Weyers, D. Kracht, "Passively Q-switched microchip laser based picosecond light source in the visible-red to near-infrared band for semiconductor excitation," Opt. Express **30**, 15428-15435 (2022)

Watt-level femtosecond Tm:(Lu,Sc)$_2$O$_3$ ceramic laser

Ning Zhang[1,2], Zhanxin Wang[1], Shande Liu[2], Wei Jing[3], Hui Huang[3], Zixuan Huang[1], Kangzhen Tian[1], Zhiyong Yang[1], Yongguang Zhao[1], Uwe Griebner[4], Valentin Petrov[4], and Weidong Chen[5]

1. *Jiangsu Key Laboratory of Advanced Laser Materials and Devices, Jiangsu Normal University, Xuzhou, 2211161, China*
2. *College of Electronic and Information Engineering, Shandong University of Science and Technology, Qingdao, 266590, China*
3. *Institute of Chemical Materials, China Academy of Engineering Physics, Mianyang, 621900, China*
4. *Max Born Institute for Nonlinear Optics and Short Pulse Spectroscopy, 12489 Berlin, Germany*
5. *Fujian Institute of Research on the Structure of Matter, Chinese Academy of Sciences, Fuzhou, 350002 Fujian, China*

The cubic sesquioxides (RE$_2$O$_3$, where RE = Lu, Y, Sc, or their mixture) combine superior thermo-mechanical properties, relatively low phonon energies, and broad and smooth gain spectra, making them attractive for high-average-power ultrafast lasers in the 2-μm spectral range. To date, nearly all sub-100-fs mode-locked bulk Tm-lasers with average powers exceeding 100 mW have relied on crystalline or ceramic type sesquioxides hosts [1]. With in-band pumping ($^3H_6 \rightarrow {}^3F_4$ transition in Tm^{3+}) which combines the high brightness of the single-mode fiber and a low quantum defect, an average power up to 1 W has been obtained from a Kerr-lens mode-locked Tm:Sc$_2$O$_3$ crystalline laser [2]. Nevertheless, growth of high-quality crystalline sesquioxides is still a challenge due to their high melting temperature (> 2400 °C [3]). In comparison, polycrystalline sesquioxide ceramics can be easily fabricated in the form of large size or "mixed" structures which exhibit inhomogeneous gain spectral broadening caused by compositional disorder. Here, we study the performance of a SESAM mode-locked Tm:(Lu,Sc)$_2$O$_3$ ceramic laser in-band pumped by a Raman fiber laser at 1627 nm. More than 1-W average power is achieved with a pulse duration of 280 fs at 2060 nm.

An astigmatically compensated *X*-shaped cavity with two flat chirped mirrors in the longer arm terminated by a flat-wedged output coupler (OC) was employed. Each chirped mirror provides a group delay dispersion (GDD) of ~ −125 fs^2 per bounce at 2.08 μm. The uncoated ceramic sample with dimensions of 3 × 3 × 3 mm^3 was placed at Brewster's angle between two concave (RoC = -100 mm) dichroic mirrors. The simulated laser beam radius on the ceramic was 32 μm × 57 μm in the sagittal and tangential planes, respectively. A SESAM was employed for starting and stabilizing the mode-locking, and the calculated beam radius on it was ~ 50 μm.

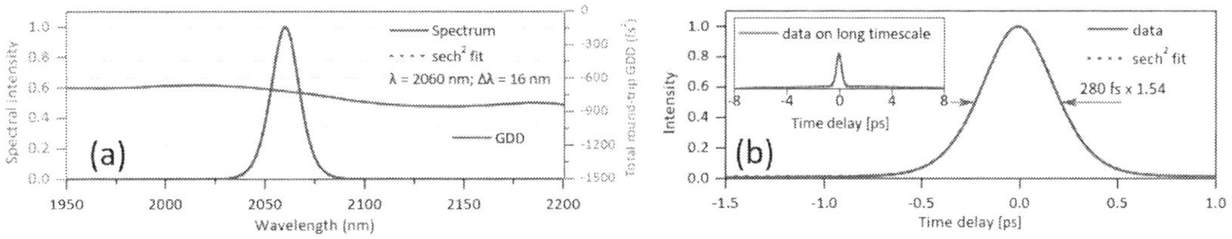

Fig. 1 Optical spectrum (a) and autocorrelation trace (b) of the SESAM mode-locked Tm:(Lu,Sc)$_2$O$_3$ ceramic laser with T_{OC} = 3%. Solid red curve in (a): total round-trip GDD. *Inset* in (b): long-timescale (± 8 ps) autocorrelation trace.

With a 3% OC, the highest average output power of 1.02 W at 86.5 MHz repetition rate was achieved from the self-starting femtosecond mode-locked Tm:(Lu,Sc)$_2$O$_3$ ceramic laser, yielding a single pulse energy of 11.8 nJ. The corresponding laser efficiency with respect to the absorbed pump power amounted to 36.4%. The measured optical spectrum, with a central wavelength of 2060 nm, is shown in Fig. 1(a). Fitting the spectral profile with a sech^2-function gives a FWHM (full width at half maximum) of 16 nm. Figure 1(b) shows the measured noncollinear autocorrelation trace at the highest average power. By assuming a sech^2 pulse intensity profile, the pulse duration amounted to 280 fs, thus leading to a time-bandwidth product (TBP) of 0.317, which is an indication of transform-limited pulses. The long-timescale autocorrelation trace shown by the inset in Fig. 4(b) is an evidence of single pulse operation of the mode-locked ceramic laser without any temporal satellites.

The demonstrated Tm:(Lu,Sc)$_2$O$_3$ ceramic laser performance is comparable to the crystalline laser in [2] and represents the highest power level ever reported from Tm- or Tm,Ho femtosecond mode-locked bulk lasers.

References

[1] V. Petrov, Y. Wang, W. Chen, Z. Pan, Y. Zhao, L. Wang, M. Mero, S. Y. Choi, F. Rotermund, W. B. Cho, W. Jing, H. Huang, H. Yuan, H. Cai, L. Zhang, Z. Lin, P. Loiko, X. Mateos, X. Xu, J. Xu, H. Yu, H. Zhang, S. Suomalainen, M. Guina, A. Härkönen, and U. Griebner, "Sub-100-fs bulk solid-state lasers near 2-micron," Proc. SPIE. **11209**, 112094G (2019).

[2] M. Tokurakawa, E. Fujita, and C. Kränkel, "Kerr-lens mode locked Tm^{3+}:Sc$_2$O$_3$ single-crystal laser in-band pumped by an Er:Yb fiber MOPA at 1611 nm," Opt. Lett. **42**(16), 3185-3188 (2016).

[3] C. Kränkel, "Rare-earth-doped sesquioxides for diode-pumped high-power lasers in the 1-, 2-, and 3-μm spectral range," IEEE J. Sel. Top. Quantum Electron. **21**(1), 1602013 (2015).

 https://doi.org/10.1051/epjconf/202226701016

Visible, femtosecond, high power, ultra-broadband noncollinear optical parametric oscillator (VIS-NOPO)

Robin Mevert[1,2], Yuliya Binhammer[1,2], Christian M. Dietrich[1,2], José R. Cardoso de Andrade[3], Luise Beichert[1,2], Thomas Binhammer[4], Jintao Fan[1,2], Uwe Morgner[1,2]

1. Leibniz University Hannover, Institute of Quantum Optics, Welfengarten 1, D-30167 Hannover, Germany

2. Cluster of Excellence PhoenixD, Welfengarten 1, D-30167 Hannover, Germany

3. Max-Born Insitute, Max-Born-Straße 2A, D-12489 Berlin, Germany

4. neoLASE GmbH, Hollerithallee 17, D-30419 Hannover, Germany

In the past, the most common approaches to generate tunable femtosecond laser pulses in the visible were either to generate intracavity SHG or SFG in near-infrared OPOs [1,2] or by pumping the OPO directly with the second harmonic of a Ti:sapphire laser (VIS-OPO) [3,4]. In this work, we demonstrate the full potential of the noncollinear phase matchings scheme in the visible by pumping the OPO with the third-harmonic of an Yb-fiber laser [5]. The resulting visible radiation can just be tuned by changing the cavity length.

The setup of the VIS-NOPO is shown in Fig. 1(a). With up to 3 W of output power the third-harmonic of an Yb-fiber MOPA pump source providing a repetition rate of 50.2 MHz and an output power up to 20 W at 1040 nm and a pulse duration of 250 fs, is generated in two BBO crystals by mixing the fundamental with its second harmonic. The third harmonic is focused into a Brewster-cut BBO ($\Theta = 39°$), which allows for an ultrabroadband noncollinear phase matching in the Poynting vector walk-off compensation (PVWC) phase matching scheme with a noncollinear angle of $\alpha = 4.5°$ between signal and pump. The cavity is setup as a standing wave cavity with two curved ($\varnothing = 25.4$ mm, ROC $= -150$ mm) and six plane ($\varnothing = 12.7$ mm) double chirped mirrors (DCM) with a high reflectivity in a range from 400 – 800 nm and a spectral transmission window for the pump. At the lower end of the cavity an output coupler (OC) with a reflectivity of 86 % in a range from 480 – 720 nm is placed on a linear translation stage. A BK7-substrate tilted at Brewster's angle is added to the cavity to stabilize the tuning behaviour of the VIS-NOPO. During the measurements, the cavity length is changed, while the output power and the autocorrelation trace of the output are recorded. The measured autocorrelation trace is fitted with a sech^2-shape to calculate the pulse duration.

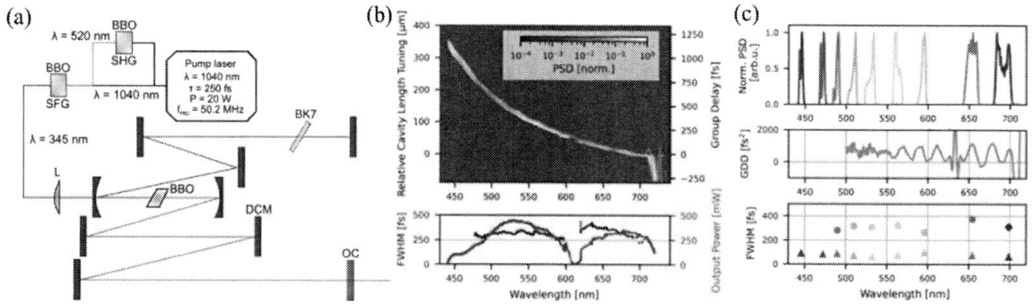

Fig. 1 (a) Setup of the VIS-NOPO: BBO SHG ($\Theta = 23.2°$, L = 1.5 mm), BBO SFG ($\Theta = 32.5°$, L = 1 mm), L: plane-convex lens (f = 200 mm), BBO OPO ($\Theta = 39°$, L = 2 mm), BK7-substrate (L = 12 mm), OC: output coupler (R = 86 %); (b) *Upper graph:* Recorded signal spectra of the VIS-NOPO while varying the cavity length at 3 W of pump power. *Lower graph:* corresponding output power (red scale on the left) and pulse duration (blue scale on the right) for center wavelengths of spectra in the upper graph. (c) *Upper graph:* arbitraly selected signal spectra of the VIS-NOPO. *Middle graph:* calculated net group delay dispersion of the VIS-NOPO cavity. *Lower graph:* measured pulse duration for selected signal spectra (circles) and calculated transform-limited pulse duration (triangles).

The tunability range of the VIS-NOPO spans from 440 – 720 nm with a maximum output power of 450 mW at 530 nm and pulse durations down to 200 fs at 528 nm, as illustrated in Fig. 1(b). The strong decrease in output power at 613 nm arises from the parasitic SHG phase matching in BBO. In conclusion, our measurements show for the first time that the VIS-NOPO can generate high power femtosecond laser pulses quickly tuneable over nearly the visible spectral range.

References

[1] Y. Khanukaeva, T. Lang, A. Tajalli, J. R. Andrade, T. Binhammer, and U. Morgner, "Efficient fast-tunable femtosecond visible radiation based on intracavity sum-frequency generation in a NIR NOPO," CLEO 2016, OSA Technical Digest (2016), paper STh1P.6

[2] T. Joeng, S. Kim, G.Kim, K.Yee, "Visible-pulse generation in gain crystal of near-infrared femtosecond optical parametric oscilator", Opt. Express **23**, pp. 25620-25627, (2015)

[3] G.M. Gale, M. Cavallari, F.Hache, "Femtosecond visible optical parametric oscillator", JOSAB **15**, pp. 702-714, (1998)

[4] M.Ghotbi, A.Esteban-Martin, M. Ebrahim-Zadeh, "BiB_3O_6 femtosecond optical parametric oscillator" Opt. Lett. **31**, pp. 3128-3130 (2006)

[5] R. Mevert, Y. Binhammer, C. M. Dietrich, L. Beichert, J. R. Cardoso de Andrade, T. Binhammer, J. Fan, and U. Morgner, "Widely tunable, high-power, femtosecond noncollinear optical parametric oscillator in the visible spectral range," Photon. Res. **9**, 1715-1718 (2021)

Sub 15 ps Self Mode-Locked Nd:YVO_4 Laser through Intra-Cavity Sum-Frequency Mixing

Martin Brunzell, Max Widarsson, Fredrik Laurell and Valdas Pasiskevicius
Department of Applied Physics, Royal Institute of Technology, Roslagstullsbacken 21, 106 91 Stockholm, Sweden

Ultrafast lasers have proven to be a great asset in many fields such as biological imaging [1], high-precision machining [2] and nonlinear spectroscopy [3]. This has resulted in a great interest in the further development of passive mode-locked sources. The most common passive mode-locking techniques today rely on semiconductor saturable absorbers (SESAM) [4] or other artificial saturable absorbers, such as Kerr lensing [5]. While these methods are well established, they still have issues with the durability of the SESAM or the need to reach sufficient intensities for reliable operation of the Kerr lens mechanism.

In this work we present a new mechanism for passively mode-locking solid-state lasers using intra-cavity sum-frequency mixing (SFM). The underlying idea is to have two laser cavities with a shared section in which a nonlinear crystal is placed. The nonlinear medium phase-matches the SFM between the two operating wavelengths. By matching the roundtrip time of the two cavities, the same temporal part of the light in the two cavities will always interact. This forces one of the lasers to form a dark pulse and the other a bright pulse. Advantages of this approach is that it works for high repetition rates, can be used for any wavelengths in the transparency window for the nonlinear material and is easy to setup. Moreover, the phase-mismatched frequency doubling in the same crystal might infer cascaded $\chi^{(2)}:\chi^{(2)}$ Kerr mechanism which could lead to spectral broadening and solitary mode-locking regime.

The setup is shown in Fig. 1, where two Nd:YVO_4 lasers resonate in a folded y-cavity, one operating at 1064 nm and the other at 1342 nm. In the shared section of the cavity a periodically poled nonlinear RKTP crystal is placed which is quasi-phase matched (QPM) for SFM between the two lasing wavelengths.

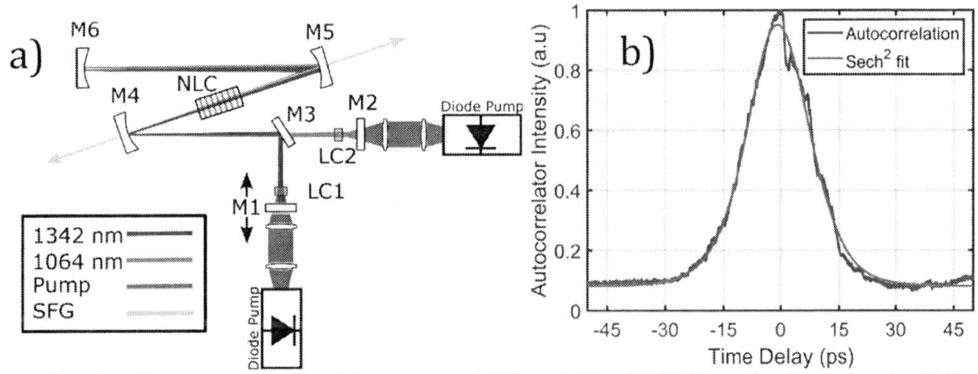

Fig. 1 a) Shows a schematic of the setup used. Where LC1 = Nd:YVO_4 with AR coating for 1342 nm, LC2 = Nd:YVO_4 with AR coating for 1064 nm, M1-M6 = mirrors for the two laser cavities, and NLC = PPRKTP with a period of 12.65 μm. Both lasers were pumped by an 808 nm laser diode. **b)** Autocorrelation of the 1064 nm pulse train, where the sech^2 fit has a FWHM of ~20 ps corresponding to a pulse width of ~13 ps.

When operating the laser, it was clear that the preferred wavelength for bright pulses was 1064 nm. One possible explanation is that in the 1064 nm laser we expect modulation instability (MI) and possible soliton formation due to action of the negative cascaded Kerr nonlinearity in the normal group velocity dispersion environment. Analysis shows that the effective Kerr nonlinearity is positive at 1342 nm. When using an output coupler of 0.4 % for the 1064 nm cavity and no output coupler for the 1342 nm cavity, a pulse width of ~13 ps was measured, see Fig.1. b. The mode-locked output power at 1064 nm was 102 mW with a repetition rate of 276 MHz.

References

[1] D. Yelin and Y. Silberberg, "Laser scanning third-harmonic-generation microscopy in biology," Opt. Express **5**, 169-175 (1999)
[2] Jian Cheng, Chang-sheng Liu, Shuo Shang, Dun Liu, Walter Perrie, Geoff Dearden, & Ken Watkins, "A review of ultrafast laser materials micromachining," Optics & Laser Technology, **46**, 88-102 (2013).
[3] A. Major, F. Yoshino, J.S. Aitchison, and P.W.E Smith, "Ultrafast nonresonant third-order optical nonlinearities in ZnSe for photonic switching at telecom wavelengths". Applied Physics Letters, **85**(20), 4606-4608 (2004).
[4] U. Keller et al., "Semiconductor saturable absorber mirrors (SESAM's) for femtosecond to nanosecond pulse generation in solid-state lasers," in IEEE Journal of Selected Topics in Quantum Electronics, **2**(3),435-453 (1996).
[5] T. Brabec, Ch. Spielmann, P. F. Curley, and F. Krausz, "Kerr lens mode locking," Opt. Lett. **17**, 1292-1294 (1992)

7.5W Alexandrite Ring Laser

Goronwy Tawy[1], Meizhen Liang[1], Huaifeng Xiao[1], Ara Minassian[2], Michael J. Damzen[1]

1. Photonics Group, The Blackett Laboratory, Dept. of Physics, Imperial College London, London SW7 2AZ, UK

2. Unilase Ltd, 60 Grays Inn Road, Unit LG04, London WC1X 8LU, UK

We report significant improvement in the performance of high-power Alexandrite ring lasers. In this work we demonstrate preliminary results for a red-diode-pumped Alexandrite laser with bidirectional output power of 7.5W with a near-diffraction limited output mode at 757nm.

Over the last decade, red-diode-pumped Alexandrite (Cr^{3+}:$BeAl_2O_4$) lasers have emerged as a highly promising laser material in the near-infrared due to its low-cost, small footprint, power scaling performance [1] as well as single-step second-harmonic-generation (SHG) to the UV-blue region. Single-longitudinal-mode (SLM) operation for applications including remote sensing and quantum technologies using diode-pumped solid-state lasers can be realised using a unidirectional ring resonator. SLM diode-pumped Alexandrite lasers have been demonstrated at the watt-level [2,3], however, for higher power operation, which is essential for efficient SHG conversion, multi-watt operation is necessary.

Fig. 1(a) shows the setup for the diode-pumped Alexandrite ring laser for bidirectional operation. Pumping is provided by a 200μm fibre-coupled red-diode at λ_p = 640nm. The pump was focused by an aspheric lens of focal length f_p = 79mm to a waist radius of w_p = 225μm. The 6mm long, 0.2 at.% Cr-doped Alexandrite crystal was mounted in a water-cooled copper mount at 20°C. The ring cavity was formed of four plane mirrors. Two dichroic mirrors (DM) were highly transmissive (HT) at λ_p and highly reflective at λ_l ~755nm, a highly reflective mirror at λ_l (HR) and an output coupler (OC) with reflectivity R_{OC} = 98%. The cavity size was set to L_x = 30mm and L_y = 25mm. A convex-mirror (CM) and quarter-waveplate (QWP) were used to retro-reflect the residual pump.

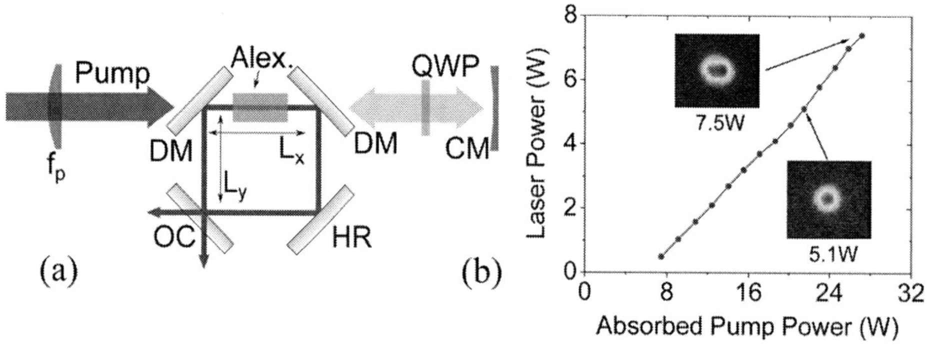

Fig. 1 (a) Diode-pumped Alexandrite ring laser with pump retro-reflection. (b) Laser power as a function of absorbed pump power for bidirectional ring cavity with mode profile at 5.1W and 7.5W of laser power.

Fig. 1(b) shows the total laser power as a function of the absorbed pump power. 7.5W was obtained at an absorbed pump power of 27.1W (28% optical efficiency) with a slope efficiency of 35%. At the maximum laser power, the beam quality was measured to be M^2 = 1.85 x 1.60, in the x-direction and y-direction, respectively. The laser wavelength was centred at 757nm with a linewidth (FWHM) of <1nm.

Better beam quality was achieved at lower pump power with M^2 < 1.1 in both directions at a laser power of 5.1W, as indicated in Fig. 1(b). The gradual degradation in beam quality above 21.5W is due to the increasing thermal lens strength reducing the TEM_{00} mode size at the gain medium. Optimal mode-matching is therefore obtained with a slightly higher-order mode. Improved beam quality should be possible with further optimisation of the cavity size.

These results show promising potential for a multi-watt wavelength tunable single-longitudinal-mode laser in the near-infrared and in the ultra-violet by means of second-harmonic-generation, which is currently in progress. Breakthrough in near-diffraction limited >10W Alexandrite lasers has also been recently demonstrated and will be presented in addition.

References

[1] G. Tawy, A. Minassian and M. J. Damzen, "High-power 7.4W TEM00 and wavelength-tunable Alexandrite laser with a novel cavity design and efficient fibre-coupled diode-pumping", OSA Continuum **3**, 1638 (2020)

[2] X. Sheng, G. Tawy, J. Sathian, A. Minassian and M. J. Damzen, "Unidirectional single-frequency operation of a continuous-wave Alexandrite ring laser with wavelength tunability", Opt. Express **26**, 31129 (2018)

[3] J. Sathian, G. Tawy, X. Sheng, A. Minassian and M. J. Damzen, "Non-astigmatic alexandrite ring laser design with wavelength-tunable single-longitudinal-mode operation", J. Opt. Soc. Am. B **37**, 2185 (2020)

Comparison of crossed-Porro prism resonator design with conventional mirror resonator design in a Ho^{3+}:YAG laser

Katharina Goth[1, 2], Michael Griesbeck[1], Madeleine Eitner[1], Marc Eichhorn[1, 2], and Christelle Kieleck[1]

1. Fraunhofer IOSB (Institute of Optronics, System Technologies and Image Exploitation), Gutleuthausstraße 1, 76275 Ettlingen, Germany

2. Institute of Control Systems, Karlsruhe Institute of Technology, Fritz-Haber-Weg 1, 76131 Karlsruhe, Germany

Ho^{3+}:YAG lasers that emit light at a wavelength of 2.09 µm have a variety of technological applications. For instance, they can be used in medicine, for atmospheric measurements, for material processing, and for the pumping of optical-parametric oscillators. Crossed Porro prisms as cavity end mirrors yield high stability against misalignments but were so far used only in four-level lasers [1]. Due to their retro-reflective property in the axis perpendicular to their apex, each of the prisms stabilizes the laser in one direction.

In this work, we present the first continuous-wave crossed-Porro-prism laser (Porro resonator) involving a Ho^{3+}:YAG crystal and compare it to a conventional resonator that uses mirrors as cavity end reflectors (mirror resonator). The crystal is pumped by a commercial 1.908 µm Tm^{3+}-doped fiber laser. Since the reflection in the Porro prisms is based on two consecutive total internal reflections and, thus, 100 % of the incident light is reflected back from the prisms, for output coupling a combination of a wave plate and a polarizer is used. Furthermore, the linear cavity contains an intra-cavity telescope, which enlarges the mode on one of the prisms, in order to reduce the intra-cavity losses.

Fig. 1(a) shows the power-power curves of the mirror and Porro resonator. While the mirror resonator shows a high slope efficiency of 74.8 % with respect to the absorbed pump power, the slope efficiency of the Porro resonator is slightly lower with a value of 67.4 %. This is attributed to losses at the apex of the prisms [2]. The maximum output power of the mirror and the Porro resonator is 37.8 W and 30.7 W, respectively. In the Porro resonator, the output power is limited by temporal instabilities that occur above an absorbed pump power of 42 W. These instabilities are eventually ascribed to instabilities in the thermal lens [3].

Fig. 1(b) shows a comparison of the beam quality of the mirror and the Porro resonator depending on the absorbed pump power. To measure the beam quality, the output beam is focused by a plano-convex lens and the $1/e^2$ beam diameter of the focused beam is measured along the propagation direction. For both resonator designs, a very good M^2 of smaller than 1.2 (with one exception) is shown. In combination with the high slope efficiency, this shows that a good mode matching between pump and laser mode is established. In the Porro resonator, the M^2 even surpasses that of the mirror resonator which is attributed to the retro-reflective property of the prisms.

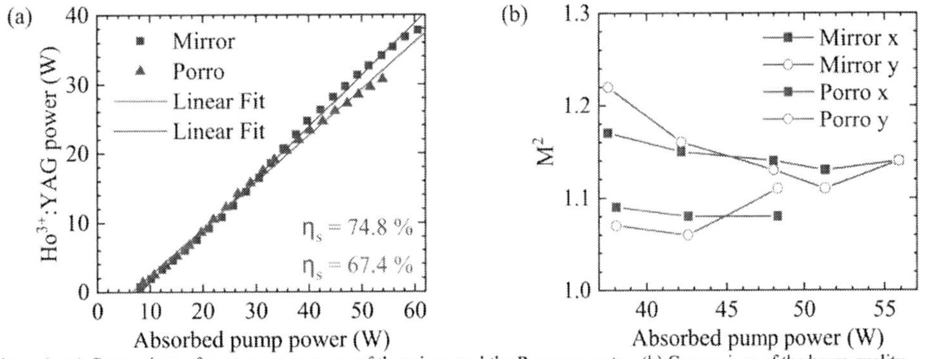

Figure 1 (a) Comparison of power-power curve of the mirror and the Porro resonator. (b) Comparison of the beam quality of the investigated resonators.

In conclusion, we presented a quasi-three-level Ho^{3+}:YAG laser involving two crossed-Porro prisms as cavity end reflectors and compared it to a conventional mirror resonator. While the Porro resonator suffered from losses at the apex of the prisms and, thus, had a reduced but still high slope efficiency in comparison to the mirror resonator, it clearly surpassed the mirror resonator in terms of beam quality while enhancing the resonator stability by a factor of up to 200 [3].

References

[1] M. K. Chun and E.A. Teppo, "Laser resonator: an electrooptically Q-switched Porro prism device", Appl. Opt. **15**, 1942-1946 (1976).
[2] I. A. Litvin, L. Burger, and A. Forbes, "Petal-like modes in Porro prism resonators", Opt. Express **15**, 14065-14077 (2007).
[3] K. Goth, M. Griesbeck, M. Eitner, H. Büker, M. Eichhorn, and C. Kieleck, "Alignment-insensitive end-pumped continuous-wave crossed-Porro prism Ho^{3+}:YAG laser", Opt. Letters (posted 01 June 2022), https://doi.org/10.1364/OL.459892 (2022).

High Power Alexandrite Laser for Tunable UV-Blue Generation

Goronwy Tawy[1,2], Noelia Palomar Davidson[2], Paolo L. Mennea[2], Glenn M. Topley[2], Peter G. R. Smith[2], James C. Gates[2], Corin B. E. Gawith[2], Ara Minassian[3], Michael J. Damzen[1]

1. Photonics Group, The Blackett Laboratory, Dept. of Physics, Imperial College London, London SW7 2AZ, UK
2. Optoelectronics Research Centre, University of Southampton, University Road, Southampton, Hampshire SO17 1BJ, UK
3. Unilase Ltd, 60 Grays Inn Road, Unit LG04, London WC1X 8LU, UK

We present, to the best of our knowledge, the first ever demonstration of tunable UV-blue generation using a diode-pumped Alexandrite laser. A high-power and narrow-linewidth red-diode-pumped Alexandrite laser operating with diffraction limited output ($M^2 < 1.1$) is frequency doubled using a BBO crystal to generate 375-385nm continuously tunable light.

Advancements in Quantum Technologies have brought about an increased demand for laser sources in the UV-blue region (360-400nm) for applications including Doppler cooling and entangled photon pair generation. Diode-pumped solid-state lasers provide excellent mode quality, linewidth and stability but with limited direct emission in the UV-blue region. Red-diode-pumped Alexandrite lasers have undergone rapid development recently with compact systems achieving multi-watt TEM_{00} operation [1]. Their broad emission at 720-800nm matches the UV-blue region via second-harmonic-generation (SHG).

Fig. 1(a) shows the setup for the system. The Alexandrite laser is pumped by a 17W fibre-coupled red-diode at $\lambda_p = 640$nm focused to a waist radius of $w_p = 225\mu$m. The laser cavity is formed of: a dichroic curved mirror of curvature R = 300mm (convex) with high transmission (HT) coating at λ_p and highly reflective (HR) at the lasing wavelength λ_l=720-800nm; and an output coupler (OC) with reflectivity $R_{OC} = 99\%$. The cavity length is around 60mm. The 4x4x6mm c-cut Alexandrite crystal with 0.2at.% Cr-doping was mounted in a water-cooled copper mount at 15°C. Wavelength tuning was obtained by rotating a 1mm thick intra-cavity birefringent filter (BiFi). SHG testing was performed by placing a BBO crystal (4x4x10mm) for Type I critical phase matching ($\theta = 30.8°$ and $\varphi = 90°$) at the laser output. The laser mode size at the OC was measured to be around 150μm.

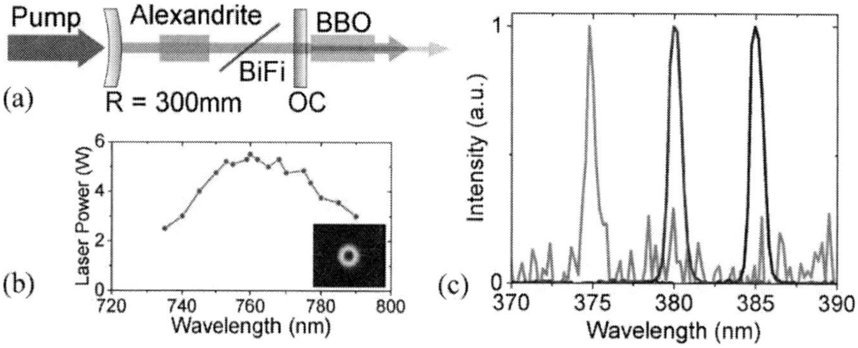

Fig. 1 (a) Wavelength tunable, diode-pumped Alexandrite laser and single-pass SHG generation using BBO. (b) Laser power as a function of wavelength. (c) SHG laser spectra at 375, 380 and 385nm.

Fig. 1(b) shows the laser power as a function of the wavelength. Continuous wavelength tuning at 735-790nm was obtained with >2.5W of laser power across the entire tuning range, and a maximum power of 5.5W at 760nm (17W absorbed pump power). The tuning was limited by the 57nm free-spectral-range of the BiFi. The beam quality was measured to be $M^2 < 1.1$ across the entire tuning range with a measured laser linewidth (FWHM), limited by the resolution of the spectrometer, of 0.2nm. SHG measurement were optimum at around 760nm with ~1mW of output power measured at 380nm. 375-385nm continuous tuning was obtained by rotating the BiFi and angular rotation of the BBO crystal. Fig. 1(c) shows the spectra at 375, 380 and 385nm.

Current work is aimed at improving the power levels and efficiency using a quasi-phase-matched material such as periodically poled lithium-niobate (PPLN) which has high nonlinearity and no spatial walk-off. PPLN waveguides have recently become highly favourable due to their high conversion efficiency, small size and low input power requirements [2]. The implementation of PPLN waveguides with red-diode-pumped Alexandrite lasers for a low-cost, compact and efficient UV-blue laser source is currently in progress and will be presented at the conference.

References

[1] G. Tawy, A. Minassian and M. J. Damzen, "High-power 7.4W TEM00 and wavelength-tunable Alexandrite laser with a novel cavity design and efficient fibre-coupled diode-pumping", OSA Continuum 3, 1638 (2020)
[2] A. C. Gray, J. R. C. Woods, L. G. Carpenter, H.Kahle, S. A. Berry, A. C. Tropper, M. Guina, V. Apostolopoulos, P. G. R. Smith, and C. B. E. Gawith, "Zinc-indiffused MgO:PPLN waveguides for blue/UV generation via VECSEL pumping," Appl. Opt. **59**, 4921 (2020)

Efficient XUV out-coupling mechanisms for intra-oscillator HHG

J. Drs,[1] J. Fischer,[1] M. Müller,[1] N. Modsching,[1] V. J. Wittwer,[1] T. Südmeyer[1]
[1]Laboratoire Temps-Fréquence, Institut de Physique, Université de Neuchâtel, Avenue de Bellevaux 51, 2000 Neuchâtel, Switzerland

High harmonic generation (HHG) using GW-level peak power ultrafast lasers enables coherent XUV light generation in laboratory-scale setups. Historically, most of these systems were based on Ti:sapphire laser amplifiers operating at kilohertz repetition rate. Nowadays, Yb-based laser systems deliver comparable peak powers, but at megahertz repetition rate, thus allowing to further increase the XUV flux. Towards even higher repetition rates, passive femtosecond enhancement cavities (fsEC) are used. The field enhancement within the cavity enables generating the required peak power even at up to several-hundred-megahertz repetition rate [1]. A simplified approach of fsEC is to drive the HHG directly within a high-power femtosecond laser oscillator [2]. Both these concepts typically use a collinear generation geometry, which requires optical elements to separate the generated XUV from the driving laser. In contrast to fsEC, the intra-oscillator approach allows for higher intra-cavity losses, thus, offering a wider choice of XUV out-coupling methods and different trade-offs between the fundamental and XUV wavelengths. In this study, we report on two of the XUV out-coupling methods, namely a coated grazing-incidence plate (GIP) and a pierced mirror, as schematically shown in Fig. 1a,b). Both these methods are highly suitable for intra-oscillator HHG since they strongly benefit from the higher tolerance to IR losses.

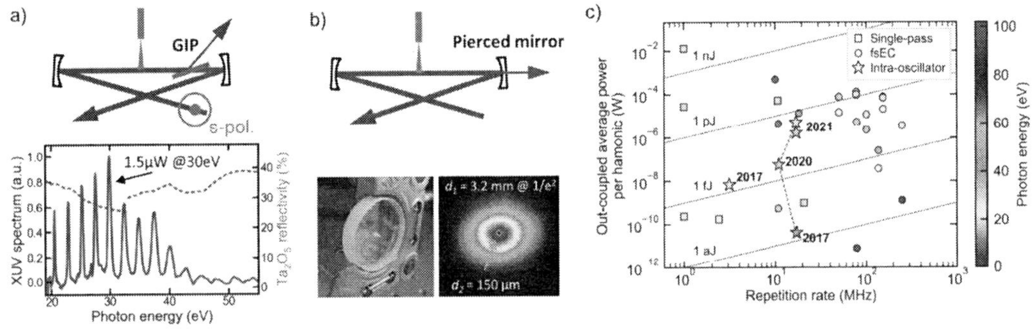

Fig. 1 a) Out-coupling scheme with a coated GIP, the measured XUV spectrum generated in Argon, and the XUV reflectivity. b) Out-coupling scheme with a pierced mirror, a photo of the pierced mirror, and the beam profile of the mode-locked laser on the mirror. d_1 – beam diameter, d_2 – hole diameter c) Overview of out-coupled XUV flux of state-of-the-art HHG systems. Stars connected with a dotted line depicts the progress of our intra-oscillator source. Our most recent results approach fsEC sources operating at similar photon energy.

We experimentally demonstrate for the first time XUV out-coupling from a cavity using a coated GIP. The concept was theoretically proposed in 2011 [3] offering high and broadband XUV reflectivity. Nevertheless, due to the high losses for the fundamental laser wavelength of up to several percent it has never been used inside an fsEC. Here we show that this concept is well-suited for intra-oscillator HHG and photon energies beyond 50 eV. The out-coupled XUV flux generated in argon amounts to 1.3 μW at 37 eV, which is approaching the results obtained with fsEC operating at similar photon energies as shown in Fig. 1c).

For even higher photon energies, a so-called pierced mirror with a small on-axis hole can be used. Likewise, this concept benefits from the higher tolerance to IR losses within a laser oscillator, allowing to increase the through-hole diameter boosting the XUV out-coupling efficiency. As a first step in this direction, we show that our thin-disk laser oscillator operates with the pierced mirror inside the cavity without a significant change in performance. Since we only have the pierced mirror on a flat substrate, we cannot yet demonstrate the XUV out-coupling due to spatial constraints of the gas target focus. The achieved intracavity performance of 1 GW at 90 fs pulse duration, however, is certainly sufficient to drive the HHG. The expected peak intensity in the gas target should allow for HHG in neon extending to photon energies even beyond 100 eV.

In conclusion, we have shown that intra-oscillator HHG systems offer new avenues for efficient XUV out-coupling thanks to their higher loss tolerance compared to fsECs. We have shown for the first time XUV cavity out-coupling using a coated GIP. In the near future, we plan to use a pierced mirror on a curved substrate which should allow us to demonstrate 100 eV photon energies by driving HHG in neon.

References

[1] I. Pupeza, C. Zhang, M. Högner, and J. Ye, "Extreme-ultraviolet frequency combs for precision metrology and attosecond science," Nat. Photonics **15**, 175 (2021).
[2] F. Labaye, M. Gaponenko, V. J. Wittwer, A. Diebold, C. Paradis, N. Modsching, L. Merceron, F. Emaury, I. J. Graumann, C. R. Phillips, C. J. Saraceno, C. Kränkel, U. Keller, and T. Südmeyer, "Extreme ultraviolet light source at a megahertz repetition rate based on high-harmonic generation inside a mode-locked thin-disk laser oscillator," Opt. Lett. **42**, 5170 (2017).
[3] O. Pronin, V. Pervak, E. Fill, J. Rauschenberger, F. Krausz, and A. Apolonski, "Ultrabroadband efficient intracavity XUV output coupler," Opt. Express **19**, 10232 (2011).

Multi-μJ 12 μm Femtosecond GaSe-based OPCPA at 1 kHz Repetition Rate

Pia Fuertjes, Martin Bock, Lorenz von Grafenstein, Uwe Griebner*, and Thomas Elsaesser

Max Born Institute for Nonlinear Optics and Short Pulse Spectroscopy, Max-Born-Str. 2a, Berlin, Germany D-12489
griebner@mbi-berlin.de

Powerful ultrashort laser sources in the long-wave infrared (LWIR) spectral region are required for new applications in strong field physics, time-resolved vibrational spectroscopy of intra- and intermolecular low-frequency modes, surface plasmons, or strongly correlated systems. In particular, amplification systems in the wavelength range beyond 10 μm are limited by the availability and transparency of nonlinear crystals and there are only a few reports of femtosecond pulse generation exceeding 1 μJ energy [1,2]. Here we present a novel table-top LWIR OPCPA working at a 1 kHz repetition rate (ν_{rep}). It contains a femtosecond Cr:ZnS oscillator as front-end providing the seed for the 2-μm Ho:YLF pump and the signal at 2.4 μm without nonlinear conversion processes. In a 3-stage GaSe OPCPA the generation of sub-200 fs idler pulses at 11.4 μm with 50 μJ energy is demonstrated.

The setup of the fs Cr:ZnS laser front-end has been reported in ref. [3]. The emission spectrum of the Cr:ZnS laser (ν_{rep}=79 MHz), extending from 1.9 – 2.6 μm, is split at 2.1 μm by a dichroic mirror and the higher frequency part is used for seeding the pump. The 2-μm Ho:YLF regenerative amplifier pump delivers compressed pulses of a 3 ps duration and 12 mJ energy at ν_{rep}=1 kHz. The 46 fs signal pulses at 2.4 μm are stretched to 0.8 ps and subsequently phase-shaped by an acousto-optic programmable dispersive filter (Fastlite). For stretching the signal and compressing the idler pulses sapphire and ZnSe crystals are used, respectively.

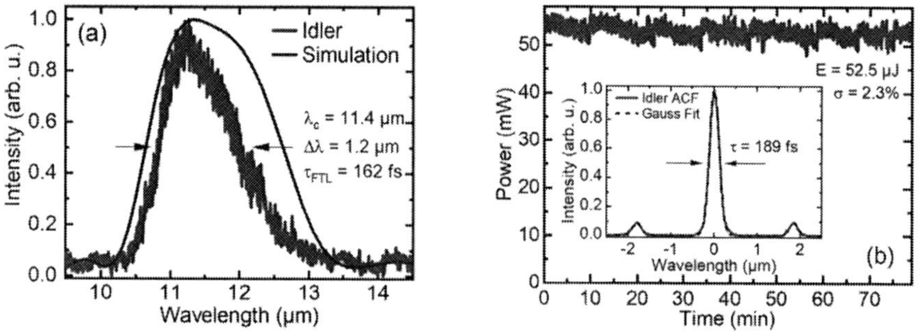

Fig. 1 Characterization of the idler pulses after the 3rd stage. (a) Measured and calculated spectrum (The latter is based on the signal spectrum after the 2nd stage.). (b) Long-term power stability; Inset: measured autocorrelation trace.

The LWIR OPA consists of 3 stages using GaSe as nonlinear crystals. In GaSe, the parametric process is phase-matched at an internal angle of 12.2° (Type-II). Employing overall 7 mJ of pump energy the combined signal and idler energy in the 1 kHz pulse train amounts to 310 μJ. The idler pulses are centered at 11.4 μm and contain 52.5 μJ energy. The resulting pump-to-idler energy conversion efficiency amounts to remarkable 0.75%. Figure 1a shows the idler spectrum after the third stage and the idler spectrum calculated with the measured signal spectrum after the second stage. A 1.2 μm bandwidth (FWHM) is achieved supporting a Fourier-transform limited (FTL) pulse duration of 162 fs. The bandwidth is limited by the phase-matching conditions for parametric frequency conversion in GaSe. Since the GaSe crystals are not anti-reflection coated, the high frequency modulation of the spectrum is associated with the 1 mm crystal thickness in the 2nd and 3rd stage. The measured autocorrelation trace of the re-compressed idler pulses is presented in the inset of Fig. 1b. Best fit to the data is achieved assuming a Gaussian-pulse shape and a duration of 189 fs is derived which corresponds to less than five optical cycles. The accompanying weak satellites are related to the modulations in the spectrum and carry about 5% of the energy. The estimated energy in the main peak amounts to 50 μJ which represents a record value for femtosecond laser sources beyond 10 μm [1,2]. The beam profile is nearly diffraction-limited with a remarkable stability of the recompressed pulses (rms: <2.5%). The latter is confirmed by a long-term power measurement shown in Fig. 1b.

[1] M. Duda, L. von Grafenstein, M. Bock, D. Ueberschaer, P. Fuertjes, L. Roškot, M. Smrž, O. Novák, and U. Griebner, "10-μJ few-cycle 12-μm source based on difference-frequency generation driven by a 1-kHz mid-wave infrared OPCPA," Opt. Lett. **47**, 2891 (2022).

[2] R. Budriunas, K. Jurkus, M. Vengris, and A. Varanavicius, "Long seed, short pump: converting Yb-doped laser radiation to multi-μJ few-cycle pulses tunable through 2.5–15 μm," Opt. Express **30**, 13009 (2022).

[3] P. Fuertjes, L. von Grafenstein, D. Ueberschaer, C. Mei, U. Griebner, and T. Elsaesser, "Compact OPCPA system seeded by a Cr:ZnS laser for generating tunable femtosecond pulses in the MWIR," Opt. Lett. **46**, 1704 (2021).

Compact cryogenic Tm:LiYF$_4$ laser

Adrian Alles,[1,2,*] Venkatesan Jambunathan,[3] Sami Slimi[2], Josep M. Serres,[1,2] Magdalena Aguiló[2], Francesc Díaz[2], Xavier Mateos,[2] Martin Smrz,[3] and Tomas Mocek[3]

[1]*Eurecat, Centre Tecnològic de Catalunya, Advanced Manufacturing Systems Unit (AMS), Marcel·lí Domingo 2, 43007 Tarragona, Spain*
[2]*Física i Cristal·lografia de Materials (FiCMA), Universitat Rovira i Virgili, c/Marcel·lí Domingo, 1, 43007 Tarragona, Spain*
[3]*HiLASE Center, Institute of Physics of the Czech Academy of Sciences, Za Radnicí 828, 25241 Dolní Břežany, Czech Republic*
**Corresponding author: adrian.alles@eurecat.org*

Lasers operating around 2 μm in the spectral range emit in the so-called eye-safe region and find applications in medicine, material processing, laser radar and atmospheric monitoring and research as well as defense. Such 2 μm lasers are achieved by doping the laser host materials with Tm^{3+} ions (here after Tm). Concerning the host, we considered LiYF$_4$ (YLF), which belongs to the fluoride family that exhibits low phonon energy, broad emission bands and long radiative lifetime, which make these materials interesting candidates for high power laser development in the 2 μm region. To date, many results based on Tm:YLF have been reported mainly at room temperature [1] and references therein. However, at room temperature reabsorption losses exist due to the quasi-three level laser nature of these systems, limiting the overall output power and efficiency. These drawbacks can be mitigated when the active medium is cooled down to cryogenic temperatures. The cryogenic cooling will result in enhancing the material thermal properties leading to improved laser performance. Cryogenic laser based on Tm:YLF has been demonstrated using broad band pumping in one of our previous work using an L-shaped asymmetric cavity. A maximum output power of 2.55 W with a slope efficiency of 22.8% was achieved [2]. In the present work, to have a compact laser setup and to enhance the laser output, we studied cryogenic laser operation of Tm:YLF using a modular setup increasing the output power above 6 W with excellent beam quality. The laser experiments were carried out using a compact modular cryogenic vacuum chamber, which has the provision to mount the optics and sample as close as possible to form a compact laser cavity. As active medium, an a- cut 5 at.% Tm:YLF crystal with 2 mm in thickness and 5×5 mm^2 aperture was used. The uncoated crystal was mounted in a copper holder at normal incidence and was placed between two plane mirrors to form a compact plane – plane cavity inside the modular chamber. As a pump source, a Volume Bragg Grating (VBG) stabilized fiber-coupled diode laser emitting at 793 nm was used and was imaged into the crystal using achromatic lenses with a ratio of 1:1.5. As output couplers, different transmissions 5, 15, and 30% were used. The cryogenic cooling of the sample was achieved by a closed cycle helium cryostat having heat load of 13W at 100K. To maintain the crystal temperature a 50 Ω heater connected to the cold finger was used.

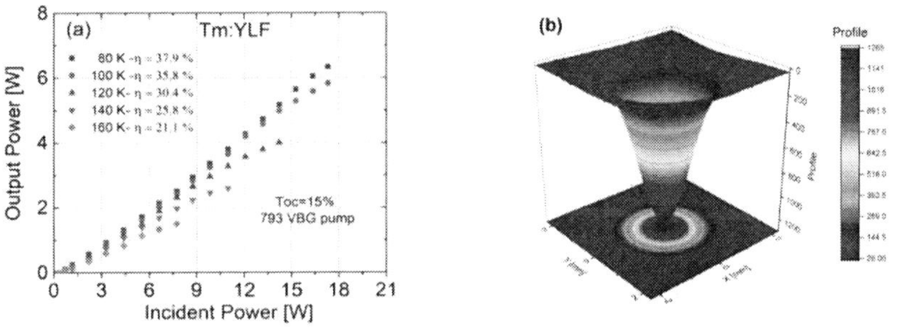

Fig. 1 (a) Input - Output power characteristics of 5at.% Tm:YLF for different temperatures, T_{OC} = 15% ; (b) Measured beam profile at maximum output power

The cryogenic continuous-wave (CW) laser was characterized by varying one of the following parameters: 1) the heat-sink temperature of the pump, 2) the output coupler transmission and 3) the temperature of the sample. In all the cases, one parameter was varied and the other two parameters remained fixed. Figure 1 (a) shows the input–output power characteristics of Tm:YLF for different temperatures with Toc=15%. From the figure, one can infer that the output power increases and laser threshold decreases with decrease in temperature. A maximum output power of 6.50 W with a slope efficiency of almost 40% at 80 K was achieved. Figure 1(b) shows the measured beam profile at maximum power, which shows an excellent Gaussian fitting as a result of the cryogenic temperature. It is worth to note, that we observed contribution of both polarizations (π and σ). At low output power levels (< 2 W), the π polarization was dominant and emits at 1876 nm and at high output power levels (>4 W), the σ polarization was dominant and emitting at 1901 & 1912 nm. All the experimental details along with the obtained results will be presented. Currently, we are focusing on pulsed laser operation using Cr:ZnS and Cr:ZnSe as saturable absorbers inside the cavity.

References

[1] P. Loiko, J. M. Serres, X. Mateos, S. Tacchini, M. Tonelli, S. Veronesi,, D. Parisi, A. Di Lieto, K. Yumashev, U. Griebner, and V. Petrov, "Comparative spectroscopic and thermo-optic study of Tm:LiLnF$_4$ (Ln = Y, Gd, and Lu) crystals for highly-efficient microchip lasers at ~2 μm," Opt. Mater. Express 7, 844-854 (2017).
[2] F. Yue, V. Jambunathan, S. P. David, J. Reiter, J. Körner, D. Klöpfel, J. Hein, M. C. Kaluza, A. Lucianetti, and T. Mocek "Diode-pumped cryogenic Tm:LiYF$_4$ laser", Proc. SPIE 11033, 110330E-1 - 110330E-6 (2019).

Sub-30 fs Kerr-lens mode-locked Ytterbium-activated orthoaluminate laser

Weidong Chen[1,2], Zhang-Lang Lin[1], Huang-Jun Zeng[1], Wen-Ze Xue[1], Ge Zhang[1], Xiaodong Xu[3], Yongguang Zhao[3], Pavel Loiko[4], Xavier Mateos[5], Haifeng Lin[6], Li Wang[2] and Valentin Petrov[2]

1. *Fujian Institute of Research on the Structure of Matter, Chinese Academy of Sciences, 350002 Fuzhou, China*
2. *Max Born Institute for Nonlinear Optics and Short Pulse Spectroscopy, Max-Born-Str. 2a, 12489 Berlin, Germany*
3. *Jiangsu Key Laboratory of Advanced Laser Materials and Devices, Jiangsu Normal University, 221116 Xuzhou, China*
4. *Centre de Recherche sur les Ions, les Matériaux et la Photonique (CIMAP), UMR 6252 CEA-CNRS-ENSICAEN, Université de Caen, 6 Boulevard du Maréchal Juin, 14050 Caen Cedex 4, France*
5. *Universitat Rovira i Virgili, URV, Física i Cristal·lografia de Materials (FiCMA), Marcel·lí Domingo 1, 43007 Tarragona, Spain*
6. *College of Physics and Optoelectronic Engineering, Shenzhen University, 518118 Shenzhen, China*

Ytterbium (Yb^{3+})-doped yttrium orthoaluminate, Yb:$YAlO_3$ (abbreviated Yb:YAP), belonging to the orthorhombic class (sp. gr. *Pnma*, perovskite-type structure), is very promising for the development of high-power femtosecond lasers at ~1 µm due to a combination of excellent thermo-mechanical and spectroscopic properties [1, 2]. Yb:YAP is optically biaxial and its intrinsic birefringence provides polarization anisotropy leading to a linearly polarized laser emission, which suppresses the thermally induced depolarization losses inherent to isotropic laser gain media. The first demonstration of femtosecond passive mode-locking of a Yb:YAP laser was reported in 2008. A SEmiconductor Saturable Absorber Mirror (SESAM) mode-locked (ML) Yb:YAP laser delivered 225 fs pulses at 1041 nm with an average output power of 0.8 W [3]. Subsequently, power scaling of the SESAM mode-locked Yb:YAP laser was achieved by using a novel off-axis pumping scheme yielding 140 fs pulses at 1009.7 nm with an average output power of 4 W [4]. Here, we report on a sub-30 fs Yb:YAP laser delivering soliton pulses as short as 24 fs at 1085 nm with an average output power of 186 mW and a pulse repetition rate of 87.5 MHz via soft-aperture Kerr-lens mode-locking (KLM).

KLM operation was investigated in an X-folded astigmatically compensated standing-wave cavity with a 5 at.% Yb^{3+}-doped YAP crystal. The uncoated sample was cut along the ***a***-axis and mounted in a water-cooled copper holder placed at Brewster's angle between two concave dichroic folding mirrors with a radius of curvature (RoC) of -100 mm. The pump source was a spatially single-mode Yb-fiber laser emitting linearly polarized output at 979 nm. The intracavity group delay dispersion (GDD) was optimized by two flat dispersive mirrors (DMs) with a total round-trip negative GDD of -1200 fs^2. The physical cavity length was ~1.7 m corresponding to a pulse repetition rate of ~87.5 MHz. The laser polarization was along the ***b***-axis.

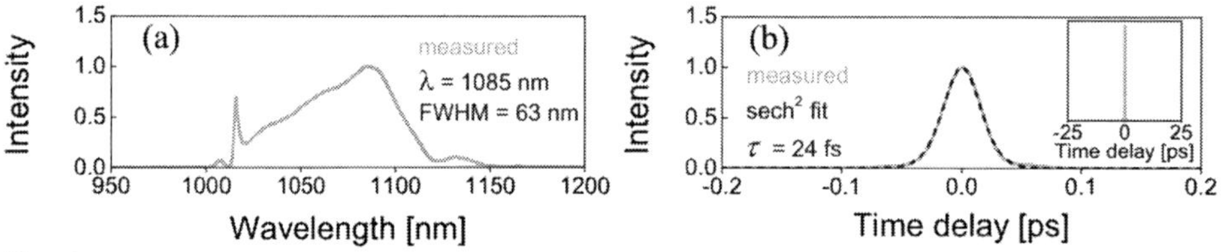

Fig. 1 KLM Yb:$YAlO_3$ laser with T_{OC} = 1%: (a) Optical spectrum and (b) SHG-based intensity autocorrelation trace. Inset in (b): simultaneously measured long-scale (50 ps) autocorrelation trace.

Soliton pulses as short as 24 fs were generated from the KLM Yb:YAP laser for 1% output coupler transmission. The maximum average output power amounted to 186 mW for an incident pump power of 2.5 W. The output spectrum was centered at 1085 nm and had a bandwidth of 63 nm (full width at half maximum, FWHM), which corresponded to a time-bandwidth product (TBP) of 0.385. The measured optical spectrum together with the corresponding second-harmonic generation (SHG) based intensity autocorrelation traces are shown in Fig. 1. The chirp could not be fully eliminated by the intracavity DMs, as can be deduced from the TBP which is larger than the Fourier-transform-limit for a $sech^2$-shaped pulse (0.315). Our result represents the shortest pulses ever achieved with any ytterbium-activated orthoaluminate crystal and highlights the potential for high-power few-optical-cycle pulse generation from a mode-locked Yb:YAP laser.

References

[1] W.-Z. Xue, Z.-L. Lin, H.-J. Zeng, G. Zhang, P. Zhang, Z. Chen, Z. Li, V. Petrov, P. Loiko, X. Mateos, H. Lin, Y. Zhao, L. Wang, and W. Chen, "Diode-pumped SESAM mode-locked Yb:(Y,Gd)AlO_3 laser," Opt. Express **30**(7), 11825-11832 (2022).
[2] R. Aggarwal, D. Ripin, J. Ochoa, and T. Fan, "Measurement of thermo-optic properties of $Y_3Al_5O_{12}$, $Lu_3Al_5O_{12}$, $YAlO_3$, $LiYF_4$, $LiLuF_4$, BaY_2F_8, $KGd(WO_4)_2$, and $KY(WO_4)_2$ laser crystals in the 80 – 300 K temperature range," J. Appl. Phys. **98**(10), 103514 (2005).
[3] V. E. Kisel, S. V. Kurilchik, A. S. Yasukevich, S. V. Grigoriev, S. A. Smirnova, and N. V. Kuleshov, "Spectroscopy and femtosecond laser performance of Yb^{3+}:$YAlO_3$ crystal," Opt. Lett. **33**(19), 2194 (2008).
[4] A. Rudenkov, V. Kisel, A. Yasukevich, K. Hovhannesyan, A. Petrosyan, and N. Kuleshov, "High power SESAM mode-locked laser based on Yb^{3+}:$YAlO_3$ bulk crystal," Devices and Methods of Measurements **11**(3), 179 (2020).

Sub-40 fs Kerr-lens mode-locked Tm,Ho:CALGO laser

Weidong Chen[1,2], Li Wang[1], Uwe Griebner[1], Ge Zhang[2], Pavel Loiko[3], Xavier Mateos[4], Ji Eun Bae[5], Fabian Rotermund[5], Xiaodong Xu[6], Arkady Major[7], and Valentin Petrov[1]

1. *Max Born Institute for Nonlinear Optics and Short Pulse Spectroscopy, Max-Born-Str. 2a, 12489 Berlin, Germany*
2. *Fujian Institute of Research on the Structure of Matter, Chinese Academy of Sciences, 350002 Fuzhou, China*
3. *Centre de Recherche sur les Ions, les Matériaux et la Photonique (CIMAP), UMR 6252 CEA-CNRS-ENSICAEN, Université de Caen, 6 Boulevard du Maréchal Juin, 14050 Caen Cedex 4, France*
4. *Universitat Rovira i Virgili, URV, Física i Cristal·lografia de Materials (FiCMA), Marcel·lí Domingo 1, 43007 Tarragona, Spain*
5. *Department of Physics, Korea Advanced Institute of Science and Technology (KAIST), 34141 Daejeon, Korea*
6. *Jiangsu Key Laboratory of Advanced Laser Materials and Devices, Jiangsu Normal University, 221116 Xuzhou, China*
7. *University of Manitoba, 75A Chancellors Circle, Winnipeg, MB, R3T 5V6, Canada*

Tetragonal calcium rare-earth (RE) aluminates, $CaREAlO_4$, where RE = Gd or Y (abbreviated CALGO and CALYO), represent a laser host crystal family characterized by strong inhomogeneous spectral line broadening due to the pronounced structural disorder when doped with laser-active RE^{3+} ions (e.g., Yb^{3+}, Tm^{3+} and/or Ho^{3+}), which is favorable for sub-100-fs pulse generation from mode-locked (ML) lasers. Despite the disordered structure, they feature relatively high thermal conductivity with moderate dependence on the RE^{3+} doping level [1], which is advantageous for power scaling when pumped with high-power diode lasers. Tm^{3+} and Ho^{3+} co-doped $CaREAlO_4$ crystals exhibit extremely broad, flat and smooth spectral gain profiles extending beyond 2 µm which originate from overlapping gain profiles of the two ions. In the first sub-100-fs ML laser of this type, a Tm,Ho:CALYO crystal was employed. Mode-locked by a SEmiconductor Saturable Absorber Mirror (SESAM), this laser delivered 87-fs pulses at 2043 nm with an average output power of only 27 mW [2]. Subsequently, a shorter pulse duration of 52 fs was achieved from a SESAM ML Tm,Ho:CALGO laser at 2015 nm with an average output power of 376 mW [3]. Very recently, sub-50-fs pulses at ~2 µm were reported using a compositionally "mixed" Tm,Ho:Ca(Gd,Lu)AlO_4 crystal (Tm,Ho:CALGLO) as a gain medium. Pulses as short as 46 fs were achieved by the SESAM ML Tm,Ho:CALGLO laser at 2033 nm with an average output power of 121 mW [4]. In the present work, we explored further shortening of the pulse duration in ML Tm,Ho:CALGO laser via soft-aperture Kerr-lens mode-locking (KLM). A transmission-type single-walled carbon-nanotube saturable absorber (SWCNT-SA) was implemented to stabilize the KLM.

The shortest pulses of the Tm,Ho:CALGO laser were achieved with a 0.2% output coupler (OC) through strong bleaching of the SWCNT-SA. KLM operation was not self-starting but required a slight perturbation. The total intracavity round-trip group-delay dispersion of the four dispersive mirrors (DMs) used amounted to -1375 fs^2. The pulses were characterized by second-harmonic generation (SHG) frequency-resolved optical gating (FROG).

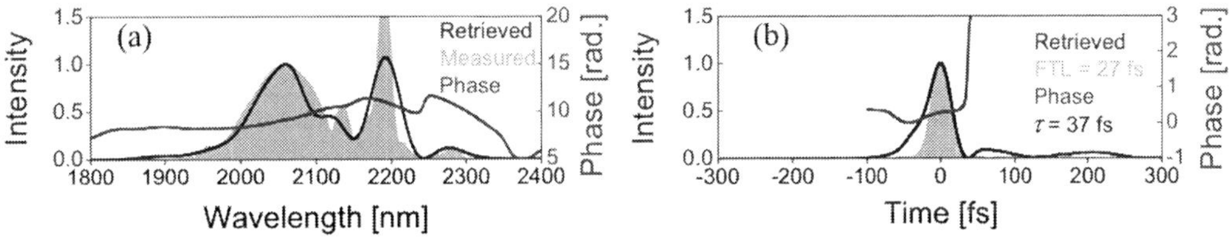

Fig. 1 SHG-FROG characterization of the KLM Tm,Ho:CALGO laser with a 0.2% OC. (a) Measured and retrieved optical spectrum and spectral phase; (b) retrieved intensity and phase and Fourier-transform-limited (FTL) pulse.

The directly measured spectrum was centered at 2061.3 nm spanning from 1870 to 2337 nm, in reasonably good agreement with the one reconstructed from the SHG-FROG trace, see Fig. 1(a). After an external linear chirp compensation with a 5-mm thick ZnSe plate, the retrieved pulse had a duration (FWHM) of 37 fs (~5 optical cycles), still ~36% above the Fourier limit, see Fig. 1(b). The average output power amounted to 55 mW at an absorbed pump power of 2.85 W and a pulse repetition rate of ~76 MHz.

References

[1] P. Loiko, F. Druon, P. Georges, B. Viana, and K. Yumashev, "Thermo-optic characterization of Yb:$CaGdAlO_4$ laser crystal," Opt. Mater. Express **4**(11), 2241 (2014).
[2] Y. Zhao, Y. Wang, X. Zhang, X. Mateos, Z. Pan, P. Loiko, W. Zhou, X. Xu, J. Xu, D. Shen, S. Suomalainen, A. Härkönen, M. Guina, U. Griebner, and V. Petrov, "87 fs mode-locked Tm,Ho:$CaYAlO_4$ laser at ~2043 nm," Opt. Lett. **43**(4), 915 (2018).
[3] V. Petrov, Y. Wang, W. Chen, Z. Pan, Y. Zhao, L. Wang, M. Mero, S. Y. Choi, F. Rotermund, W. B. Cho, W. Jing, H. Huang, H. Yuan, H. Cai, L. Zhang, Z. Lin, P. Loiko, X. Mateos, X. Xu, J. Xu, H. Yu, H. Zhang, S. Suomalainen, M. Guina, A. Härkönen, and U. Griebner, "Sub-100-fs bulk solid-state lasers near 2-micron," Proc. SPIE. **11209**, 112094G (2019).
[4] L. Wang, W. Chen, Y. Zhao, P. Loiko, X. Mateos, M. Guina, Z. Pan, M. Mero, U. Griebner, and V. Petrov, "Sub-50 fs pulse generation from a SESAM mode-locked Tm,Ho-codoped calcium aluminate laser," Opt. Lett. **46**(11), 2642 (2021).

https://doi.org/10.1051/epjconf/202226701026

VCSELs as Highly Sensitive Stand-Alone Distance Sensors

Axel Günther [1,2,3], Divyaben Korat [2], Wolfgang Kowalsky [1,3] and Bernhard Roth [2,3]
[1] Institute of High Frequency Technology, TU Braunschweig, Schleinitzstr. 22, 38106 Braunschweig, Germany
[2] Hannover Centre for Optical Technologies, Leibniz Universität Hannover, Nienburger Str. 17, 30167 Hannover, Germany
[3] Cluster of Excellence PhoenixD (Photonics, Optics and Engineering – Innovation Across Disciplines), Hannover, Germany
Contact: axel.guenther@ihf.tu-bs.de

The development of novel optical components as alternatives to their electronic counterparts has become a very active field of research during the last decades [1]. Especially for sensing purposes a wide variety of different applications has been investigated to measure surface topographies, temperature changes or mechanical strain, among others [2]. Next to the investigation of optical waveguiding structures also the field of light sources as key components for optical communication or miniaturized optical measurement devices, respectively, showed significant advance. Especially vertical-cavity surface-emitting lasers (VCSEL) have become important components for data transmission in data centers [3,4].
In our work we extend the field of application for the VCSEL by using them as light sources and sensing elements simultaneously. This allows highly sensitive detection of surface topographies of integrated optical structures opening the path towards their use as inline and online monitoring tool in integrated optics production [5].
We investigate a novel sensor concept based on a coupled resonator configuration and the use of VCSELs as central elements. The back reflection of a sample surface next to the emission window of the VCSEL affects the internal resonator conditions of the laser resulting in a change of the emitted wavelength and operating current. The cross section in figure 1 a) depicts the fundamental parts of a VCSEL. The two distributed Bragg reflection (DBR) grating stacks with the active layer located in between them form the internal resonator. Next to the emission window on the upper part a partially reflecting sample is placed. The light which is reflected back into the VCSEL interferes with the internal resonator modes. This leads to a slight shift of the emission wavelength, figure 1 b), and of the operating current, respectively, when the operation voltage is kept constant. The formed coupled resonator is very sensitive to movements of the sample along the optical axis of the emitted laser beam.

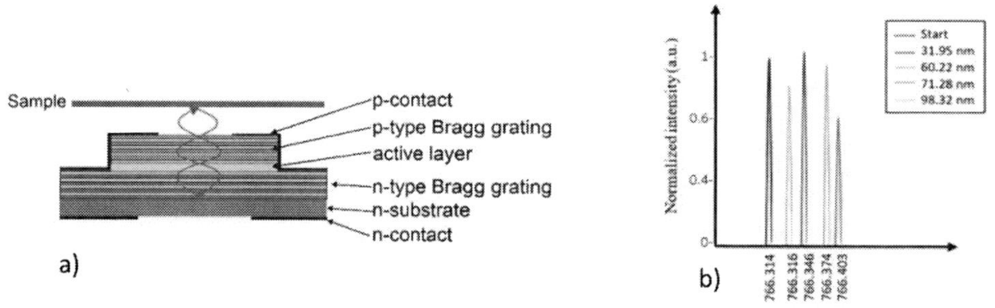

Fig. 1: Cross section of a standard VCSEL showing the active layer and the distributed Bragg reflection (DBR) grating which are essential parts of the developed system. The sample next to the emission window reflects the light partially back into the resonator where it interacts with the internal modes forming a coupled resonator configuration.

The setting enables us to detect changes of the distance between sample and emission window of a few nm only with high precision. In future, we will integrate the new sensor systems into an inline production process chain for additive manufacturing of optics to verify the topography of the created structures and establish a closed feedback loop in for correction of manufacturing inaccuracies.

References

[1] N.K. Guimard, N. Gomez and C.E. Schmidt, "Conducting polymers in biomedical engineering", Prog. Polymer Sci. **32**, 876-921 (2007)
[2] H. Tazawa, T. Kanie and M. Katayama, „Fiber-optic coupler based refractive index sensor and ist application to biosensing", Appl Phys. Lett. **92**, 113901 (2007)
[3] J. Tatum, D. Gazula, L. Graham, J. Guenter, R. Johnson, J. King, C. Kocot, G. Landry, I. Lyubomirsky, A. MacInnes, E. Shaw, K. Balemarthy, R. Shubochkin, D. Vaidya, M. Yan and F. Tang, "VCSEL-based interconnects for current and furture data centers", J. Lightwave Technol. **33**, 727-732 (2015)
[4] G. Kanakis, N. Iliadis, W. Soenen, B. Moeneclaey, N. Aryris, D. Kalavrouziotis, S. Spiga, P. Bakopoulos and H. Avramopoulos, "High-speed vcsel-based transceiver for 200 Gb short-reach intra-datacenter optical interconnects", Appl. Sci **9**, 2488 (2019)
[5] C. Gorecki, S. Khalfallah, H. Kawakatsu and Y. Arakawa, "New SNOM sensor using optical feedback in a VCSEL-based compound-cavity", Sens. Actuator A Phys. **87**, 113-123 (2001)

The impact of heat-load modulation on transverse mode instability in high-power, quasi-continuous wave fibre amplifiers

Sobhy Kholaif[1], Cesar Jauregui[1], *Yiming Tu[1,2]* , *Jens Limpert[1,2,3]*

1. *Instiute of Applied Physics, Abbe Center of Photonics, Friedrich-Schiller-Universität Jena, Albert-Einstein-Str. 15, 07745 Jena, Germany*
2. *Helmholtz-Institute Jena, Fröbelstieg 3, 07743 Jena, Germany*
3. *Fraunhofer Institute for Applied Optics and Precision Engineering, Albert-Einstein-Str. 7, 07745 Jena, Germany*

Nowadays, transverse mode instability (TMI) constitutes the main limitation for a further power scaling of fibre laser systems with diffraction-limited beam quality [1]. TMI is observed when a certain average output power level (i.e., the TMI threshold) is reached. Above this level a dynamic deterioration of the beam profile suddenly occurs [2]. The root cause of this effect is that the modal interference pattern (MIP) in the fibre core- between the fundamental mode (FM) and at least one higher-order mode (HOM)- induces a phase-matched refractive-index grating (RIG) through the thermo-optic effect. This RIG allows for a modal energy transfer between the fibre modes if there is a phase-shift between the MIP and the RIG. This is because the sign of this phase shift determines the direction of the modal energy transfer. In a previous theoretical study, it has been revealed that a permanent energy transfer from the HOMs to the FM can be induced above the TMI threshold when pulsing the input seed with some specific modulation parameters [3]. In this way, the heat load is modulated, which allows controlling the sign of the phase shift.

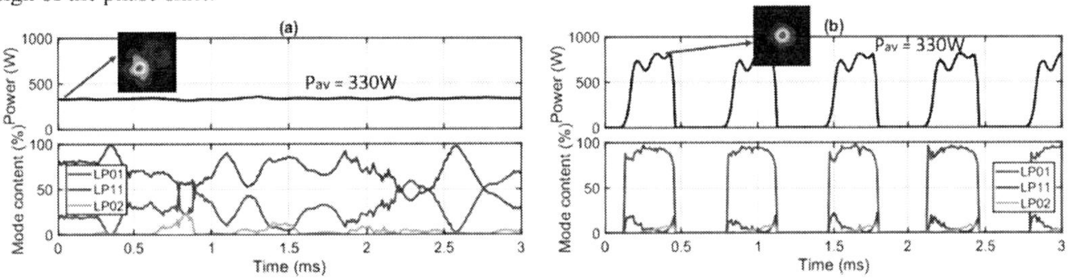

Fig. 1 Modal analysis at an average output power of 330 W for **a)** free-running system and **b)** heat load modulation technique by pulsing the seed and pump with a modulation frequency of 1.5 kHz, a duty cycle of 45 %, 65 % TMI increase, and dominant FM time-window of 300 μs. The insets are intensity frames measured by HSC with 10 μs temporal resolution.

In this work, we experimentally demonstrate dominant FM operation at an average output power significantly higher than the TMI threshold via heat load modulation [4]. The heat-load modulation can be achieved by pulsing the seed, the pump or both simultaneously. With this technique it is possible to obtain a beam which energy is predominantly contained in the FM at all times even at an average output power of 370 W, which is 75% higher than the TMI threshold of the free-running system. This result is even more impressive considering that the instantaneous output power of the pulsed signal is higher than 850 W (i.e., over 4.1 times higher than the TMI threshold). Figure 1 shows an example of this measurement to compare between the modal content analysis [5], at the same average output power of 330 W, for a free-running system (figure 1a) and the heat-load modulation (figure 1b) at a modulation frequency of 1.5 kHz, a duty cycle of 45 % and a modulation depth of 100 %. The upper frames describe the temporal evolution of the instantaneous output signal power, whereas the lower frames describe the temporal evolution of the relative FM/HOMs modal content. As can be seen, the FM is dominant in the on-time interval of the modulated period (i.e., 300 μs). This result is beneficial in improving quasi-continuous wave (QCW) fibre laser systems which are vital for a variety of applications [6].

References

1. C. Jauregui, C. Stihler, and J. Limpert, "Transverse mode instability," Adv. Opt. Photonics **12**, 429 (2020).
2. T. Eidam, C. Wirth, C. Jauregui, F. Stutzki, F. Jansen, H.-J. Otto, O. Schmidt, T. Schreiber, J. Limpert, and A. Tünnermann, "Experimental observations of the threshold-like onset of mode instabilities in high power fiber amplifiers," Opt. Express **19**, 13218 (2011).
3. C. Jauregui-Misas, C. Stihler, J. Limpert, and A. Tünnermann, "Transverse mode instabilities in burst operation of high-power fiber laser systems,", Proc. SPIE 10512, 6 (2018).
4. S. E. Kholaif, C. Jáuregui-Misas, Y. Tu, and J. Limpert, "Mitigation of transverse mode instability by heat load modulation in high-power fiber laser amplifiers," Proc. SPIE 11981, 57 (2022).
5. F. Stutzki, H.-J. Otto, F. Jansen, C. Gaida, C. Jauregui, J. Limpert, and A. Tünnermann, "High-speed modal decomposition of mode instabilities in high-power fiber lasers," Opt. Lett. **36**, 4572 (2011).
6. Marco Mendes, "Fiber laser micromachining in high-volume manufacturing," Ind. Laser Solut. **1**, (2015).

 https://doi.org/10.1051/epjconf/202226701028

577 nm yellow laser source using external pumping

Great Chayran, [1,*] **Venkatesan Jambunathan,**[1] **Martin Smrz,**[1] **and Tomas Mocek**[1]
[1]*HiLASE Center, Institute of Physics of the Czech Academy of Sciences, Za Radnicí 828, 25241 Dolní Břežany, Czech Republic*
**Corresponding author:chayran@fzu.cz*

Solid-state laser has found applications in retinal photocoagulation treatment for a long time now. However, it has been proven that a laser source at 577 nm serves better than the currently used 532 nm laser sources [1]. It's due to higher absorption in oxyhemoglobin and lower macular xanthophyll absorption at 577 nm than 532 nm hence lower energy is required to treat the eye, which in turn leads to a reduced recovery period after surgery. But due to the unavailability of a 577 nm laser source, a 532 nm laser source has been adopted despite not being optimum. In our previous work, we demonstrated how to achieve a yellow laser source at 577 nm with a proper combination of gain, Raman, and frequency doubling media using intracavity pumping [2]. As an alternative approach, here in this work, we employed external pumping to generate 577 nm.

The 577 nm laser cavity was a simple hemispherical cavity with a set of optics and also it contains a Raman medium and frequency doubling medium inside the cavity. This Raman resonator was pumped using an in-house developed external pump source, which is an electro-optically Q- Switched cryogenic Yb: YAG laser emitting around 1029 nm and generates pulse energy up to 1.4 mJ with a rep-rate of 1 KHz and Pulse width of 30 ns [3]. To collimate and focus the pump beam to the Raman resonator several other optical elements were used that includes optical isolator to avoid feedback light. The focal spot on the Raman medium was estimated to be around 130 µm in radius using 150mm focal length lens. The laser mode radius on the Raman cavity was estimated to be around 140 µm. As Raman medium, Barium Nitrate ($Ba(NO_3)_2$) crystal of length 20 mm with 5*5 mm^2 aperture was used to convert the fundamental wavelength to Raman wavelength (1154 nm). As a frequency doubling medium, three different Lithium triborate (hereafter LBO) crystals of lengths 10, 20, and 30 mm with 3*3 mm^2 aperture were used to convert to 577 nm. As output couplers, Toc=90 and 95% with Roc =100 mm coated for yellow wavelength were used.

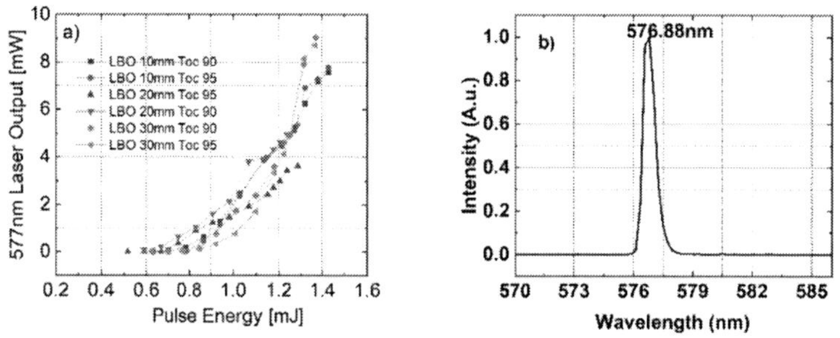

Fig. 1 (a) 577nm yellow laser output characteristics using $Ba(NO_3)_2$, different LBO's and Toc = 90 and 95%; (b) Measured yellow laser spectrum.

577 nm yellow laser was generated for all the combinations used. Figure 1 (a) presents the 577 nm laser output vs input pulse energy corresponding to the combinations of different LBO's, Toc's = 90 and 95% and 20 mm $Ba(NO_3)_2$. From the results, one can infer that the output power does not vary much with the change in transmission of output coupling. In addition, one can observe the output power increases non-linearly with the increase in incident energy. To be noted, better phase matching and good beam quality were obtained only with 10 mm LBO crystal, although maximum output power of 9 mW was obtained with 30 mm LBO. We believe, that with longer LBO's (20 mm and 30 mm) the mode matching becomes non-uniform near the LBO's, which results in poor beam quality. Figure 1 (b) shows the measured yellow nm laser spectrum output of around 577 nm. Present work is focused on using output coupling mirrors with longer Roc (150 mm and 200 mm) for better mode matching and trails with longer Raman medium (50mm) to enhance the output power.

References

[1] G. Staurenghi, "Why consider the yellow 577 nm wavelength", Retina Today (July/August), 31-33 (2016).
[2] V Jambunathan et al, "Development of a Yellow Laser Source at 577 nm for Ophthalmology Applications ", 2021 Conference on Lasers and Electro-Optics Europe & European Quantum Electronics Conference (CLEO/Europe-EQEC), 2021, pp. 1-1,
[3] V Jambunathan et al, "Efficient laser performance of a cryogenic Yb: YAG laser pumped by fiber-coupled 940 and 969 nm laser diodes", Laser Phys. Lett. 12 015002 (2015)

https://doi.org/10.1051/epjconf/202226701029

Multi-mJ SWIR OPCPA pumped and seeded with 1.2 ps Yb:YAG laser

Augustinas Petrulėnas[1], Aistė Butkutė[1], Paulius Mackonis[1] and Aleksej M. Rodin[1]
1. Solid State Laser laboratory, Center for Physical Sciences and Technology, Savanoriu 231, LT-02300 Vilnius, Lithuania

Few-cycle short-wavelength infrared (SWIR) pulses are in demand for research in strong field physics and nonlinear optics, efficient generation of attosecond X-ray pulses and THz radiation. At the same time, broadband SWIR-MIR pulses, covering many important molecular vibrations, are useful for remote sensing and LIDAR. Lasers with multi-mJ femtosecond pulses at a wavelength of 2 μm are needed to pump Mid-IR parametric amplifiers based on non-oxide crystals (ZGP, $CdSiP_2$, OP-GaAs) to go beyond 4 μm. Optical parametric chirped pulse amplification (OPCPA) has been successfully applied to generate multi-mJ SWIR pulses. One possible implementation of the SWIR OPCPA is based on the use of a small portion of the pump laser energy to generate a supercontinuum (SC) up to 2.5 μm, which is amplified in one or more OPCPA stages. We demonstrate a non-collinear amplification of white-light supercontinuum seed pulses in three degenerate OPCPA (DOPA) stages based on bismuth triborate (BiBO) crystals up to 2.2 mJ in the wavelength range of 1.9 – 2.3 μm.

The experimental setup of the three-stage SWIR OPCPA is shown in Fig. 1. The pump source for both SC and OPCPA is a homemade two-stage double-pass Yb:YAG chirped pulse amplifier with a grating compressor that provides transform-limited ~1.2 ps pulses with an output energy of up to 20 mJ at a repetition rate of 100 Hz [1]. SC seed pulses were generated in a YAG crystal 15 mm long [2] and stretched to 550 fs in 30 mm long ZnSe plate. Amplification of broadband seed pulses was carried out in a three-stage noncollinear OPCPA based on BiBO crystals (type-I, phase matching angle $\theta = 8°$) AR-coated in the range of 1750 – 2500 nm. The sizes of BiBO crystals in three OPCPA stages are 4×4×6 mm³ for OPA 1, 10×10×4 mm³ for OPA 2 and 15×15×4 mm³ for OPA 3, respectively. After amplification, the pulses were compressed in a Suprasil 300 glass plate of 55 mm long.

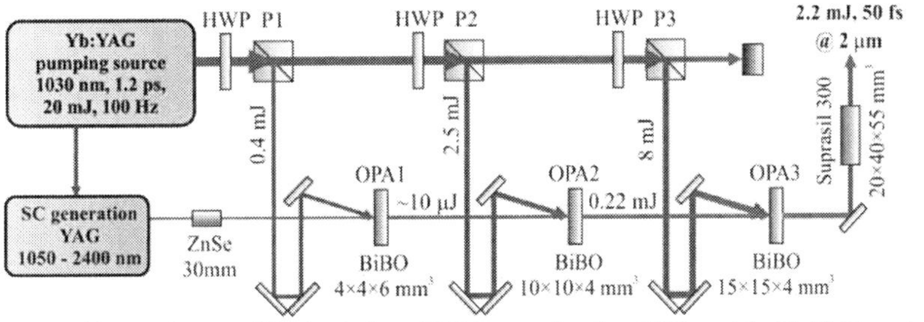

Fig. 1 Experimental layout of SC amplification in three OPCPA stages based on BiBO crystals: P1, P2, P3 –polarizing cubes, HWP – half wave plate, ZnSe– zinc selenide plate.

In the first OPCPA stage, the best conversion efficiency of ~3% with maximum signal energy of >10 μJ was observed at a pump intensity of 60 GW/cm². In the second OPCPA stage, pump-to-signal conversion efficiency of 10% was achieved at a pump intensity of 40 GW/cm². After two OPCPA stages, the signal amplified to 220 μJ. Finally, in the third OPCPA stage achieved a record conversion efficiency of 30% and output pulses with energies greater than 2 mJ at a center wavelength of 2.05 μm, supporting a transform-limited pulse width of ~30 fs. An even higher output energy can be obtained by using the maximum pump energy in BiBO crystals with a larger aperture. After three OPCPA staged, the pulses were compressed from 480 fs to 50 fs in a glass plate.

Thus, we have developed SWIR OPCPA with output pulses of 40 GW peak power in the 1.9 – 2.3 μm wavelength range. This is sufficient for filamentation in gas cells filled by N_2 or H_2 and a significant expansion of the spectrum to wavelengths beyond 3 μm due to the excitation of rotational stimulated Raman scattering.

This research was sponsored by Research Council of Lithuania under contract S-MIP-21-30. Two authors of the work are also grateful for partial financial support from NATO Science for Peace and Security Programme under grant G5734.

References

[1] P. Mackonis and A.M. Rodin, "Laser with 1.2 ps, 20 mJ pulses at 100 Hz based on CPA with a low doping level Yb: YAG rods for seeding and pumping of OPCPA," Opt. Express **28(2)** 1261-1268(2020).
[2] P. Mackonis, A. Petrulėnas, V. Girdauskas and A.M. Rodin, "Stable 1100-2400 nm supercontinuum in YAG with picosecond pumping for simplified OPCPA,"In the European Conference on Lasers and Electro-Optics (p. ca_p_43). Optical Society of America, (2019).

Recent Progress in Laser Crystals and Ceramics for Femtosecond Mode-Locked Lasers at ~2 μm [Invited]

Pavel Loiko,[1,*] Weidong Chen,[2] Xavier Mateos,[3,#] Patrice Camy,[1] Uwe Griebner,[4] and Valentin Petrov[4]

[1]*Centre de Recherche sur les Ions, les Matériaux et la Photonique (CIMAP), UMR 6252 CEA-CNRS-ENSICAEN, Université de Caen, 6 Boulevard Maréchal Juin, 14050 Caen Cedex 4, France*
[2]*Fujian Institute of Research on the Structure of Matter, Chinese Academy of Sciences, Fuzhou, 350002 Fujian, China*
[3]*FiCMA-FiCNA-EMaS, Universitat Rovira i Virgili (URV), Campus Sescelades, E-43007 Tarragona, Spain. #Serra Húnter Fellow*
[4]*Max Born Institute for Nonlinear Optics and Short Pulse Spectroscopy, Max-Born-Str. 2a, 12489 Berlin, Germany*
*pavel.loiko@ensicaen.fr

Lasers emitting femtosecond (fs) pulses in the spectral range of ~2 μm are of interest for development of amplified systems for high harmonic and THz generation, as well as pumping of mid-infrared ultrafast optical parametric oscillators based on non-oxide nonlinear crystals. Such emission can be generated using thulium (Tm^{3+}, $^3F_4 \rightarrow {}^3H_6$) and holmium ($Ho^{3+}$, $^5I_7 \rightarrow {}^5I_8$) ions. From the material point of view, for reaching short (sub-100 fs) pulse durations, there exist two challenges: (i) the need to operate above 2 μm to avoid the structured water vapor absorption in the atmosphere leading to ps pulse durations and (ii) the need of laser-active media with broad, flat and smooth gain profiles supporting the generation of ultrashort pulses. In the present work, we review recent advances in developing singly Tm^{3+} doped and Tm^{3+}, Ho^{3+} codoped laser crystals and transparent ceramics capable of generating sub-100 fs pulses from ~2 μm mode-locked (ML) solid-state lasers.

One solution to address the desired gain properties is the use of structurally disordered host crystals featuring a significant inhomogeneous spectral broadening for the dopant Tm^{3+} and Ho^{3+} ions. Two prominent examples: cubic calcium niobium gallium garnets ($Ca_3Nb_{1.5}Ga_{3.5}O_{12}$-type, CNGG) [1] and tetragonal calcium rare-earth aluminates ($CaGdAlO_4$-type, CALGO), Fig. 1(a) [2]. Even for singly Tm^{3+}-doped crystals, the emission bands are smooth and broad extending above 2 μm. A Tm:CLNGG laser passively ML by a single-walled carbon nanotube (SWCNT) saturable absorber generated 78 fs pulses at 2017 nm at a repetition rate of 86 MHz [1]. Via Tm^{3+}, Ho^{3+} codoping, one can further red-shift the emission wavelength and benefit from the combined gain bandwidths, Fig. 1(b). Using a GaSb-based SESAM, a Tm,Ho:CALGO ML laser delivered 52 fs at 2015 nm with a spectral bandwidth of 82 nm and an average output power of 376 mW [2]. We show that long-wave emissions at wavelengths well exceeding those of purely electronic transitions (the phonon sidebands of Tm^{3+} and Ho^{3+} emission bands) could be responsible for gain up to at least 2.3 μm, Fig. 1(b), thus supporting the generation of shorter pulses from ML lasers. Such emissions of vibronic nature are related to multiphonon-assisted transitions.

Another solution is based on employing materials with strong crystal-fields leading to large Stark splitting of the Tm^{3+} ground-state ($^3H_6$) naturally providing long-wave electronic emissions. One example: cubic sesquioxides A_2O_3 (A = Y, Lu, Sc or their mixture) being very suitable for the transparent ceramic technology. Using the ability of the ceramic sintering technology to fabricate easily compositionally "mixed" solid-solutions such as $(A_{1-x}B_x)_2O_3$, additional spectral broadening can be achieved. A Kerr-lens mode-locked (KLM) Tm:$(Lu,Sc)_2O_3$ ceramic laser generated pulses as short as 58 fs at ~2081 nm with an average output power of 220 mW at a pulse repetition rate of 84.8 MHz (extinction ratio above the noise level: 77 dBc). The emitted spectrum at the long-wavelength wing extended to 2.25 μm due to the vibronic transitions of the Tm^{3+} ions. The latter is found to be essential for generating pulses with durations in the sub-50 fs range.

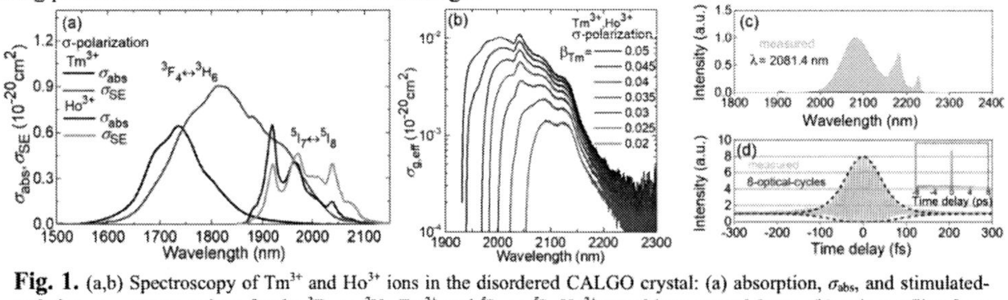

Fig. 1. (a,b) Spectroscopy of Tm^{3+} and Ho^{3+} ions in the disordered CALGO crystal: (a) absorption, σ_{abs}, and stimulated-emission, σ_{SE}, cross-sections for the $^3F_4 \leftrightarrow {}^3H_6$ Tm^{3+} and $^5I_7 \leftrightarrow {}^5I_8$ Ho^{3+} transitions around 2 μm, (b) gain profiles for a Tm^{3+}, Ho^{3+} codoped crystal plotted in a semi-log scale accounting for the long-wave vibronic emission. Light polarization: σ; (c,d) KLM Tm:$(Lu,Sc)_2O_3$ sesquioxide ceramic laser: (c) optical spectrum and (b) interferometric autocorrelation trace.

References

[1] Y. Wang, Y. Zhao, Z. Pan, J. E. Bae, S. Y. Choi, F. Rotermund, P. Loiko, J. M. Serres, X. Mateos, H. Yu, H. Zhang, M. Mero, U. Griebner, and V. Petrov, "78 fs SWCNT-SA mode-locked Tm:CLNGG disordered garnet crystal laser at 2017 nm," Opt. Lett. **43**(17), 4268-4271 (2018).
[2] Y. Wang, P. Loiko, Y. Zhao, Z. Pan, W. Chen, M. Mero, X. Xu, J. Xu, X. Mateos, A. Major, M. Guina, V. Petrov, and U. Griebner, "Polarized spectroscopy and SESAM mode-locking of Tm,Ho:CALGO," Opt. Express **30**(5), 7883-7893 (2022).
[3] Y. Zhao, L. Wang, W. Chen, P. Loiko, Y. Wang, Z. Pan, H. Yang, W. Jing, H. Huang, J. Liu, X. Mateos, Z. Wang, X. Xu, U. Griebner, and V. Petrov, "Kerr-lens mode-locked Tm-doped sesquioxide ceramic laser," Opt. Lett. **46**(14), 3428-3431 (2021).

https://doi.org/10.1051/epjconf/202226701031

Compact Nd:YAP/V:YAG nanosecond pulse generator at 1342 nm

Kryštof Kadlec[1], Jan Šulc[1], Michal Němec[1], Helena Jelínková[1], Karel Nejezchleb[2], Lukáš Beran[2], Radim Kudělka[2]

1. FNSPE, Czech Technical University in Prague, Břehová 7, 115 19 Prague, Czech Republic
2. Crytur, Ltd. Turnov, Na Lukách 2283, 511 01 Turnov, Czech Republic

The aim of this work was to design a compact passively Q-switched diode-pumped laser generating stable nanosecond pulses in the spectral region of 1.34 µm with pulse energy reaching 0.2 mJ. Such a system could serve as a compact laser source of long-range high-resolution LIDAR for autonomous vehicle control. The use of radiation in this spectral region (compared to 1 µm sources) is advantageous because the background radiation intensity is reduced due to its absorption in atmospheric water vapour, and the 1.3 µm radiation is safer for the eye due to its higher absorption in water in comparison to 1.06 µm radiation – see Fig. 1.[1, 2] As a proven compact source of short pulses at a wavelength of 1.3 µm represents a microchip laser based on the Nd:YAG/V:YAG composite.[3] Previously we demonstrated a more flexible laser system based on independent Nd:YAG and V:YAG crystals, operating at wavelength 1318 nm.[4] Currently we present similar system based on the combination Nd:YAP/V:YAG crystals, which offer better wavelength and polarization stability.

The designed laser followed our previous system based on the Nd:YAG/V:YAG combination.[4] In the current version we used the Nd:YAP crystal as a gain medium (16 mm long, 0.5 at.% Nd/Y) and a similar V:YAG saturable absorber ([100]-cut, $T_0 = 85\,\%$ @ 1.3 µm). The mirrors of the laser resonator were directly deposited on the Nd:YAP crystal (pumping mirror highly reflecting at 1.3 µm radiation; highly transparent for 808 nm) and on the V:YAG crystal (output coupler with reflection 85 % at at 1.3 µm). For longitudinal Nd:YAP pumping, a fibre-coupled laser diode (fibre core diameter 400 µm; wavelength 805 nm, $f_{rep} = 10 - 500\,\text{Hz}$, $\Delta t = 1\,\text{ms}$, $E_{pmp} = 18\,\text{mJ}$) was used. Two aspheric lenses were used to focus pumping radiation into the Nd:YAP crystal (pumping beam waist was 280 µm).

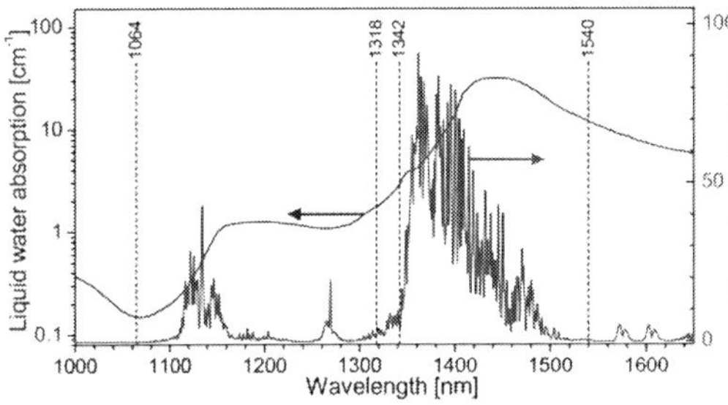

Fig. 1. Liquid water and atmospheric absorption spectra.

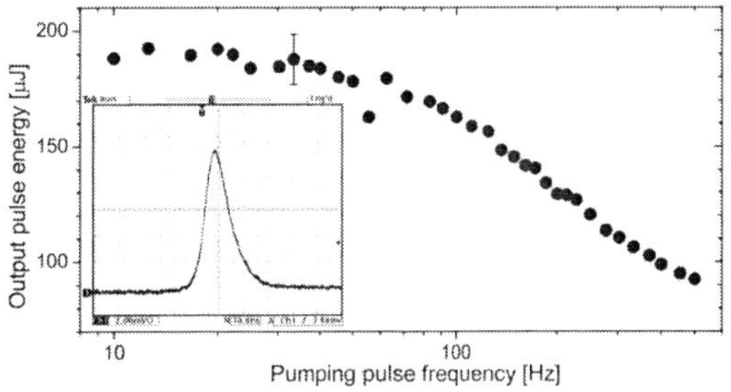

Fig. 2. Pulse energy in dependence on pumping pulse repetition rate (inset – oscillogram example).

The generated pulse energy in dependence on pumping pulse frequency is shown in Fig. 2. The maximum reached pulse energy 192 µJ corresponds to the peak power of 16 kW. In the whole range of tested frequencies, the laser generated stable ~ 12 ns long pulses (FWHM, Fig. 2 – inset) at wavelength 1342 nm. The output beam with the TEM_{00} cross-section was linearly polarized (degree of polarization > 99 %). The relative fluctuation of pulse duration, pulse energy, and pulse peak power at a fixed pumping repetition rate was lower than 1 %. We can conclude that realized compact laser source of 1.34 µm radiation is a more flexible variant of the microchip laser and due to the stability of the output radiation parameters, it is suitable both for distance measurement and for further nonlinear conversion of the generated radiation, both to visible and far infrared spectra.

References

[1] T. J. Kane, "1.34 µm Nd:YVO_4 laser passively Q-switched by V:YAG and optimized for LIDAR," in *Solid State Lasers XXIX: Technology and Devices*, W. A. Clarkson and R. K. Shori, eds., **11259**, p. 1125902, SPIE, 2020.

[2] H. Jelínková, J. Pašta, M. Němec, J. Šulc, M. Miyagi, Y.-W. Shi, Y. Matsuura, and M. Jelínek, "Different influence of long and short mid-infrared laser pulses on eye tissue," *Laser Physics, Laser methods in medicine and biology* **13**, pp. 735–742, May 2003.

[3] J. Šulc, H. Jelínková, K. Nejezchleb, and V. Škoda, "Nd:YAG/V:YAG monolithic microchip laser operating at 1.3 µm," *Optical Materials* **30**(1), pp. 50–53, 2007.

[4] J. Šulc, M. Němec, H. Jelínková, K. Nejezchleb, and Š. Uxa, "Stable 1318 nm emission from compact Nd:YAG/V:YAG Q-switched laser," in *High Power Lasers and Applications*, J. Hein, T. J. Butcher, P. Bakule, C. L. Haefner, G. Korn, and L. O. Silva, eds., **11777**, p. 117770V, SPIE, 2021.

Conversion of Mode-Locked States within an Optical Cavity

Michael Zwilich, Florian Schepers, Carsten Fallnich
University of Münster, Institute of Applied Physics, Corrensstr. 2, 48149 Münster, Germany

Locking longitudinal modes of an optical resonator results in the generation of short laser pulses, whereas, locking its transverse modes results in a beam rapidly scanning the transverse plane. Common schemes for generating such transverse mode-locked beams involve off-axis pumping of a gain medium inside a resonator [1–3]. Which modes, and to what extent, are involved in the locking process has been analyzed [1] and controlled by simple means, e.g., by limiting the highest mode order with a slit [1], transversely displacing the pump spot [2] or varying the pump power [3]. Here, we present the generation of transverse mode-locked beams with improved control over the modal amplitudes by longitudinal to transversal mode conversion. We also consider the inverse process of transverse to longitudinal mode conversion to generate temporal pulses when injecting a spatially oscillating beam.

A simple longitudinal mode-locked beam can, for example, be generated by modulating a single frequency laser beam in amplitude yielding at least two sidebands with a frequency separation $\Delta\nu_L$ to the carrier ν_0 (see Fig. 1a). However, the transverse modes of a cavity are frequency shifted by $\Delta\nu_T$ due to their specific Gouy phase shifts. Then, if the transverse mode spacing matches the modulation frequency ($\Delta\nu_T = \Delta\nu_L$), not only the carrier but also each sideband frequency can excite a separate transverse mode within the cavity and will be transmitted. As the incident longitudinal modes, that is carrier and the sidebands, are in phase with each other, so are the excited transverse modes, resulting in a transverse mode-locked output beam.

This scheme can be reversed and should result in conversion of a transverse mode-locked input, i.e. a scanning beam, to a longitudinal mode-locked output, i.e. a temporal pulsed beam.

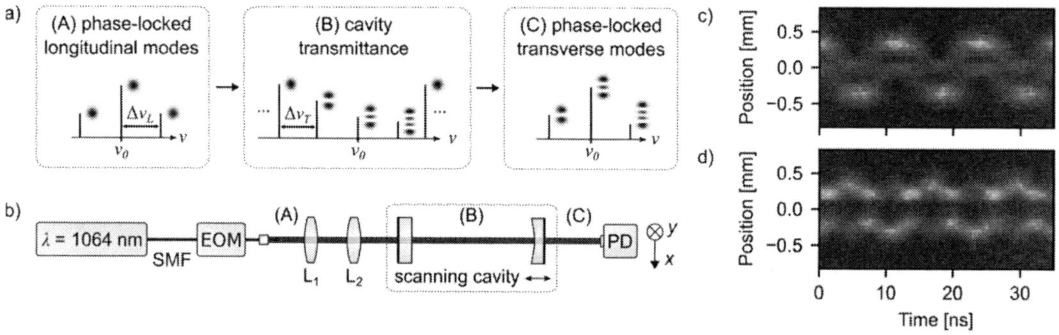

Fig. 1 Schematic of the conversion from a longitudinal mode-locked state into a transverse mode-locked state in a). Experimental realization in b). Measured resulting spatiotemporal oscillations with $\Delta\nu_T = 80$ MHz for two different sets of modal amplitudes in c) and d). Single mode fiber (SMF), electro-optic modulator (EOM), lens (L) and photodiode (PD).

For a proof-of-principle experiment, with a sinusoidal modulation signal two sidebands were imprinted on a longitudinal mode which enabled the generation of mode-locked beams comprised of a superposition of three transverse modes. The Hermite-Gaussian (HG) modes $HG_{0,1}$, $HG_{0,2}$ and $HG_{0,3}$ were brought into resonance with the incident frequencies by tuning the cavity length appropriately. The measured spatiotemporal distributions (see Fig. 1c and d) have clearly visible oscillation periods being inverse to the transverse mode spacing (1/80 MHz = 12.5 ns) characteristic for transverse mode-locked beams. In this scheme the amplitude modulation signal acting upon the output beam of a single frequency laser allows to control the amplitudes of the generated sidebands, which affects how strongly each transverse mode is excited. This made it possible to switch between a $0.24\times HG_{0,1} + 0.60\times HG_{0,2} + 0.15\times HG_{0,3}$ superposition (see Fig. 1c) to a configuration with modal amplitudes of 0.57, 0.15 and 0.28, respectively (see Fig. 1d) and thereby alter the resulting spatiotemporal oscillation.

In conclusion, we demonstrated the generation of transverse mode-locked beams by conversion from a longitudinal mode-locked beam. This allows to not only control which transverse modes are excited, but also to influence their amplitudes and the resulting spatiotemporal oscillation.

References

[1] F. Schepers, T. Hellwig, and C. Fallnich, "Modal reconstruction of transverse mode-locked laser beams," Appl. Phys. B **126**, 168 (2020).
[2] Y. Shen, Z. Wan, X. Fu, Q. Liu, and M. Gong, "Vortex lattices with transverse-mode-locking states switching in a large-aperture off-axis-pumped solid-state laser," J. Opt. Soc. Am. B **35**, 2940 (2018).
[3] S. Zhang, Z. Fu, L. Lai, F. Jia, D. Qiao, Y. Fan, K. Li, and N. Copner, "Transverse mode locking of different frequency degenerate families based on annular beam pumping," Opt. Lett. **46**, 3195 (2021).

Cryogenic Laser Operation of a "Mixed" Yb:YLuAG Garnet Crystal

Sami Slimi[1,*], Venkatesan Jambunathan[2], Mingyan Pan[3], Yicheng Wang[4], Weidong Chen[5], Pavel Loiko[6], Rosa Maria Solé[1], Magdalena Aguiló[1], Francesc Díaz[1], Martin Smrz[2], Tomas Mocek[2], and Xavier Mateos[1,#]

1.FiCMA-FiCNA-EMaS, Universitat Rovira i Virgili (URV), Campus Sescelades, E-43007 Tarragona, Spain

2.HiLASE Centre, Institute of Physics of the Czech Academy of Sciences, Za Radnici 828, 252 41 Dolní Břežany, Czech Republic

3. Key Laboratory of Materials for High Power Laser, Shanghai Institute of Optics and Fine Mechanics, Chinese Academy of Sciences, Shanghai 201800, China

4 Photonics and Ultrafast Laser Science, Ruhr Universität Bochum, Universitätsstrasse 150, 44801 Bochum, Germany

5 Fujian Institute of Research on the Structure of Matter, Chinese Academy of Sciences, Fuzhou, 350002 Fujian, China

6. Centre de Recherche sur les Ions, les Matériaux et la Photonique (CIMAP), UMR 6252 CEA-CNRS-ENSICAEN, Université de Caen Normandie, 6 Boulevard Maréchal Juin, 14050 Caen Cedex 4, France

**e-mail : sami.slimi@urv.cat, #Serra Húnter Fellow*

In recent years, solid state lasers and based on Ytterbium (Yb^{3+}) ions emitting at ~1 μm have attracted significant attention due to their vast technological and scientific applications. At room temperature, Yb^{3+} ions represent a quasi-three-level laser scheme because of the finite population of Stark sub-levels of the ground state ($^2F_{7/2}$), leading to reabsorption at the laser wavelength that increases the laser threshold and hinders the optical efficiency. To overcome this limitation, the active medium could be cooled down to cryogenic temperatures, approaching the quasi-four-level laser scheme leading to much lower laser thresholds [1]. In addition, this helps to improve the thermal and thermo-optical properties of the gain material. Yttrium aluminium garnet $Y_3Al_5O_{12}$ (YAG) is the state-of-the-art laser host crystal for Yb^{3+} doping. In the present work, we report on the cryogenic laser operation of a compositionally "mixed" $(Y,Lu)_3Al_5O_{12}$ (YLuAG) garnet doped with Yb^{3+} ions.

A single crystal of Yb:YLuAG was grown by the Czochralski method using an [111] oriented seed in an Ar atmosphere. According to the ICP-AES analysis, its composition was $(Y_{0.554}Lu_{0.280}Yb_{0.166})_3Al_5O_{12}$. The crystal possessed a cubic garnet structure. The selected "mixed" composition is motived by: (i) lower melting point as compared to $Lu_3Al_5O_{12}$, (ii) lower consumption of the Lu_2O_3 reagent, (iii) additional inhomogeneous broadening of Yb^{3+} spectral bands [2].

The cryogenic laser experiments were carried out using an L-shaped asymmetric cavity. The uncoated laser element had an aperture of 4×4 mm² and a thickness of 2 mm. It was mounted in a copper holder at normal incidence and fixed to the cold finger of the optical cryostat providing a heat load of 13 W at 100 K. To maintain the sample temperature, a 50 Ω heater was connected to the cold finger and the sample temperature was monitored using a silicon sensor. As a pump source, a Volume Bragg grating (VBG) stabilized fiber-coupled diode laser emitting at 969 nm with a spectral bandwidth of 0.40 nm (N.A. = 0.22, fiber core diameter: 105 μm) was used. The unpolarized pump radiation was reimaged using achromatic lenses with a ratio of 1:2.5. For passive Q-switching, Cr^{4+}:YAG saturable absorbers (SAs) with 85% and 95% initial transmissions were used.

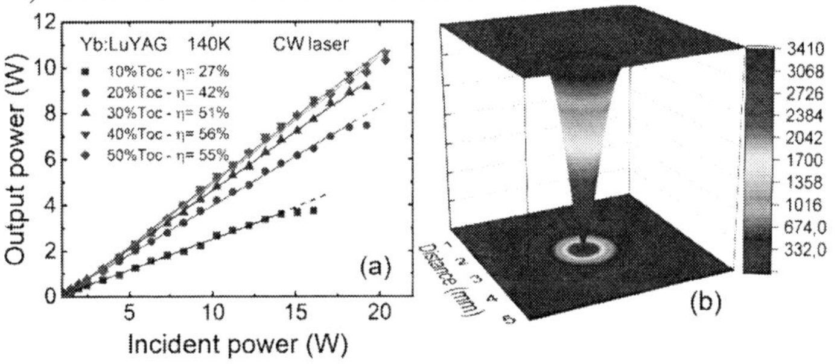

Fig. 1. CW cryogenic diode-pumped Yb:YLuAG laser: (a) input-output power characteristics for different transmissions of the output coupler (T_{OC}), η – slope efficiency, T = 140 K; (b) a typical far-field beam profile at the maximum output power.

Continuous-wave (CW) laser operation was realized initially at different temperatures in the range of 80 - 280 K with a step of 20 K using the same output coupler transmission (T_{OC} = 20%). As a result, we determined an optimum crystal temperature of 140 K for efficient laser operation, see Fig. 1(a) for the input-output dependences. The cryogenic diode-pumped Yb:YLuAG laser generated a maximum CW output power of 10.65 W at 1029 nm with a slope efficiency of 56% (with respect to the incident pump power) for T_{OC} = 40%. Figure 1(b) shows the far-field beam profile measured at the maximum output power corresponding to the fundamental transverse mode., In the passively Q-switched regime, using the SA with the initial transmission of 85%, the pulse energy / duration were 0.15 mJ / 201 ns at a repetition rate of 19.22 kHz.

References

[1]T.Y. Fan, D. J. Ripin, R. Aggarwal, J. R. Ochoa, B. Chann, M. Tilleman, and J. Spitzberg, "Cryogenic Yb^{3+}-doped solid-state laser", IEEE J. Sel. Top. Quantum Electron. **13**, 448-459 (2007).

[2] F. Wang, G. Xie, J. Xu, L. Qian, P. Yuan, X. Xu, and Z. Qin, "8.5 W mode-locked $Yb:Lu_{1.5}Y_{1.5}Al_5O_{12}$ laser with master oscillator power amplifiers", Appl. Opt. **54**, 1041–1045 (2015).

Colloidal $LiYF_4$:Pr nanocrystals downsized to 10 nm – Part 2: spectroscopic properties

Simon Spelthann[1], Michael Steinke[1,2], Rajesh Komban[3], Horst Weller[3,4], Christoph Gimmler[3], Axel Ruehl[1,2], Detlev Ristau[1,2,5]

1. Leibniz University Hannover, Institute of Quantum Optics, Welfengarten 1, D-30167, Hannover, Germany
2. Leibniz University Hannover, QUEST-Leibniz-Research School, Callinstraße 36, D-30167 Hannover, Germany
3. Fraunhofer Center for Applied Nanotechnology CAN – (a research division of Fraunhofer Institute for Applied Polymer Research, Grindelallee 117, D-20146 Hamburg, Germany
4. University Hamburg, Department of Chemistry, Grindelallee 117, D-20146 Hamburg, Germany
5. Cluster of Excellence PhoenixD (Photonics, Optics, and Engineering – Innovation Across Disciplines), Welfengarten 1A, D-30167, Hannover, Germany

Praseodymium doped fluoride crystals are well-known to be efficient laser media for the visible spectral range [1, 2]. Downsizing such crystals to the nanoscale promises additional degrees of freedom for novel composite photonic devices. However, scattering constitutes the limiting loss factor in such devices so that the size of the nanocrystals should be well below the wavelengths of the involved photons. The first part of our joint contribution is about the challenging synthesis of monodispersed colloidal $LiYF_4$:Pr nanocrystals with low doping concentrations between 0.7 and 1.47 at% . In this second part, we present the results of the extensive spectroscopic characterization of these nanocrystals. The absorption spectra are strongly broadened and exhibit fewer peaks than expected, c.f. Fig 1a). The emission spectra, however, show multiple narrow lines as expected from $LiYF_4$:Pr bulk crystals, see Fig 1b). This behavior contradicts the broadened absorption spectra because both spectra should be reciprocal according to McCumber theory. In addition, we compared the emission spectra of the nanocrystals with their bulk counterparts and observed an additional emission at 595 nm, see inset in Figure 1b). While the other visible emission occurs through spontaneous relaxation from the ^3P-manifold of the Pr-ions this emission originates from the subjacent $^1D_2$ energy level. We investigated the spontaneous emission characteristics of the ^3P-manifold. It exhibits excited state lifetimes between 15 and 18µs, which is about half of the lifetime of bulk crystals. Simultaneously, the quantum yield of the nanocrystals is below 5%, which seems to contradict their considerably high lifetimes. Surprisingly, the $^1D_2$ lifetime of the nanocrystals agrees very well with the respective lifetimes of the bulk crystals. Altogether, we observed contradicting spectroscopic behavior during the investigation of the nanocrystals. In our contribution, we will present a detailed spectroscopic investigation of these nanocrystals and discuss the role of involved quenching mechanisms to explain these contradictions. We will further present how the optical behavior of the nanocrystals' changes through downsizing to the 10nm range.

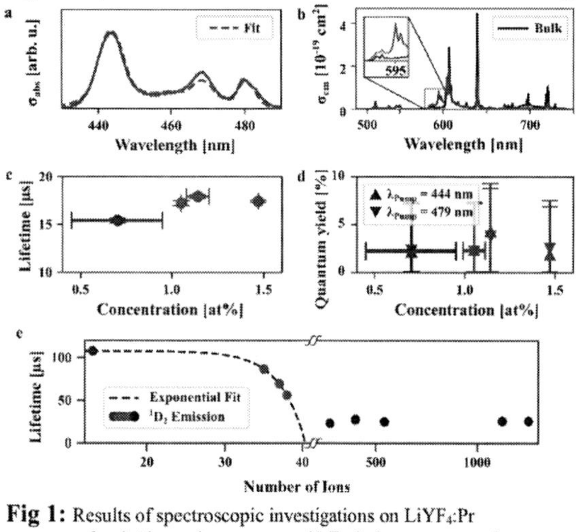

Fig 1: Results of spectroscopic investigations on $LiYF_4$:Pr nanocrystals. a) Absorption spectrum, b) Emission spectrum after excitation at 479 nm, c) Lifetimes of the $^3P_0$ level, d) Quantum yield, e) Lifetimes of the 595 nm transition after excitation at 479 nm.

Literature

[1] A. Richter, E. Heumann, E. Osiac, G. Huber, W. Seelert, A. Diening, "Diode pumping of a continuous-wave Pr^{3+}-doped $LiYF_4$ Laser", Opt. Lett. **29**, 2638 (2004)
[2] A. Richter, E. Heumann, G. Huber, V. Ostroumov, W. Seelert, "Power scaling of semiconductor laser pumped Praseodymium lasers, Opt. Exp. **15**, 5172 (2007)

EPJ Web of Conferences 267, 01035 (2022) https://doi.org/10.1051/epjconf/202226701035
EUROPHOTON 2022

10 mJ-level Picosecond OPCPA Pump Laser Based on Room Temperature Hybrid Yb:YAG Amplifier System

Aivaras Kazakevičius[1,2], Raimundas Burokas[1,2], Rokas Danilevičius[2], Andrejus Michailovas[1,2]

1. National Center for Physical Sciences and Technology, Savanorių pr. 231, LT-02300, Vilnius, Lithuania
2. Ekspla ltd., Savanorių pr. 237, LT-02300 Vilnius, Lithuania

High energy, high average power picosecond laser systems are applicable in a variety of scientific and industrial fields such as attosecond science, EUV or THz radiation generation, single shot laser marking, high speed and large area processing and many more. These lasers are also commonly used as pump sources in high intensity optical parametric chirped pulse amplification (OPCPA) systems with sub-ps or ps pulse durations being favorable due to lower amplified parametric fluorescence background in the ps to ns temporal window. End pumped Yb:YAG amplifiers feature extremely good parameters for high energy and average power laser development with major problems being nonlinear temporal and thermally induced spatial distortions. Several state of the art laser technologies have been proposed to overcome these limitations, such as thin disc [1], slab [2] and coherent beam combining [3]. However, these technologies rely on highly complex and expensive optical arrangements, which are not always necessary to avoid beam distortions.

Our aim is to create a compact and cost effective 100W-class, high energy laser system based on hybrid laser approach comprising fiber laser technology and free-space amplification combined with our patented thermal aberration management techniques [4]. During the first development stage we chose laser operation regime that allowed us to avoid polarization and beam distortions due to thermal effects and investigate amplification and nonlinear temporal pulse characteristics of the resulting laser radiation. In this work we demonstrate 1 ps laser system capable of delivering > 13 mJ energy laser radiation at 20 Hz.

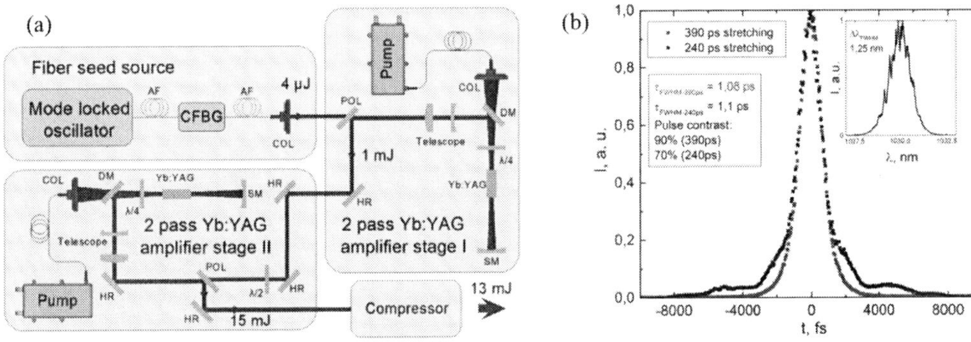

Fig. 1. (a) Schematic representation of 10 mJ, 20 Hz laser system. AF – fiber amplifier, CFBG – chirped fiber Bragg grating, λ/4, λ/2 – quarter-wave and half-wave plates, COL – beam collimator, POL – thin film polarizer, DM – dichroic mirror, SM – spherical mirror, HR – high reflectivity mirror. (b) Intensity autocorrelation of the output pulse at different pulse stretching ratios and pulse spectrum (inset).

The laser system is shown in Fig. 1. Ultrashort picosecond pulses were generated in a SESAM mode-locked fiber oscillator, stretched in a chirped fiber Bragg grating (CFBG) and amplified in several single mode fiber amplifiers resulting in 4 μJ energy pulses. Pulse energy was further increased to 1 mJ and 15 mJ in the 1st and 2nd free space Yb:YAG amplifiers. The output pulse spectrum width was ≈1.25 nm and corresponded to 1 ps Fourier transform-limited pulse duration. Pulse compression experiment showed that stretching up to 390 ps in the fiber laser was necessary to achieve 1,08 ps pulses with reasonable 90% pulse contrast (Fig. 1. b). Finally, beam quality proved to be close to diffraction limited ($M^2 \leq 1.08$) whereas the second harmonic generation experiment resulted in maximum 82% conversion efficiency indicating negligible output pulse phase distortion. These results proved that such optical arrangement is suitable to achieve high energy output pulses with desired spectral and temporal characteristics and allows us to move forward with 100W system development.

References

[1] C. Herkommer, P. Krötz, R. Jung, S. Klingebiel, C. Wandt, R. Bessing, P. Walch, T. Produit, K. Michel, D. Bauer, R. Kienberger, T. Metzger, "Ultrafast thin-disk multipass amplifier with 720 mJ operating at kilohertz repetition rate for applications in atmospheric research", Opt. Express **25**, 30164-30173 (2020).
[2] B. E. Schmidt, A. Hage, T. Mans, F. Légaré, H. J. Wörner, "Highly stable, 54 mJ Yb-InnoSlab laser platform at 0,5 kW average power", Opt. Express, **25**, 17549 – 17555 (2017).
[3] M. Müller, C. Aleshire, A. Klenke, E. Haddad, F. Légaré, A. Tünnermann, J. Limpert, "10.4 kW coherently combined ultrafast fiber laser", Opt. Lett., **45**, 3083-3086 (2020).
[4] L. Veselis, R. Burokas, O. Ulčinas, T. Gertus, K. Michailovas, A. Michailovas, "Depolarization compensation with a spatially variable wave plate in a 116W, 441 fs, 1 MHz Yb:YAG double pass laser amplifier", Appl. Opt., **60**, 7164-7171 (2021).

https://doi.org/10.1051/epjconf/202226701036

Novel coercive field engineering technique for improved periodic poling of $KTiOPO_4$ isomorphs

Laura Barrett[1], Andrius Zukauskas[1], Fredrik Laurell[1]

1. Department of Applied Physics, Royal Institute of Technology, Roslagstullsbacken 21, SE-106 91 Stockholm, Sweden

Coercive field engineering of periodically poled $KTiOPO_4$ (PPKTP) for quasi-phase matching (QPM) applications has successfully been done with Rb^+-ion indiffusion (Rb-exchange) in $KTiOPO_4$ (KTP) isomorphs [1]. To date, Rb-exchange is the only viable technique for periodic poling of bulk QPM structures with sub-μm periodicities in KTP [2]. Rb-exchange is also used to form waveguides in KTP isomorphs as it creates a refractive index increase in the exchanged layer [3]. However, for bulk QPM applications this feature is undesirable since it induces a shift in the phase-matching wavelength, which may cause unwanted spectral broadening of the generated light. Due to the large Rb^+-ion there is also a significant stress imposed in the crystal, which results in a higher fragility and lower PPKTP yield. Until now, barium has been used as an addition to the Rb-exchange recipe to increase the depth of the Rb^+ indiffusion [4]. Here we show the use of standalone Ba^{2+}-ion indiffusion as an effective method for coercive field engineering for creating short period PPKTP for QPM applications, which avoids the introduction of stress and the unwanted phase-matching wavelength shift. With this method we have poled 3.77 μm periods in flux grown KTP and Rb-doped KTP (RKTP) with record high second harmonic generation (SHG) efficiencies of 2 %/Wcm and 3.4 %/Wcm, respectively.

Similarly to Rb-exchange, Ba-exchange involves photolithographic patterning of the polar surface of the KTP crystal. A stop layer for ion-exchange is created on the crystal by O_2 plasma etching, and it is subsequently submerged for 4 hours into an exchange melt consisting of 7 At. % $Ba(NO_3)_2$ and 93 At. % KNO_3, which is kept at 375°C for the exchange duration. The crystal is then poled with 5 ms triangular voltage pulses. Figure 1(a) shows an example of how multiple voltage pulses can be used for poling a Ba-exchanged crystal, while measuring the SHG output after each pulse. The same crystal is poled with multiple pulses and then poled back to single domain in three cycles during the measurement. The results show no degradation in SHG output when re-poling the crystal in the second and third cycles. The demonstrated poling method together with possibilities for re-poling samples is useful for increasing both the quality and the yield of PPKTP.

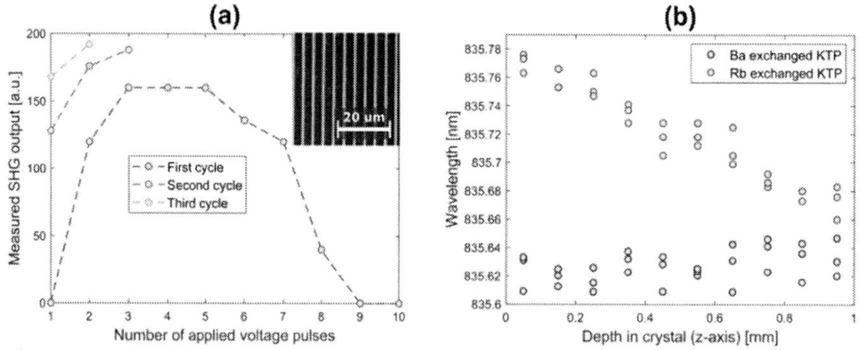

Fig. 1 (a) The measured SHG output when successive 6 kV 5 ms triangular pulses are used for poling a periodically Ba-exchanged RKTP crystal. After the first cycle the crystal is poled back to the original polarisation and the second cycle is measured, similarly the third. The inset shows a PFM image of the poled structure. (b) SHG wavelength variation as a function of the depth in Rb- and Ba-exchanged KTP crystals, where z=0 is the exchanged surface.

Figure 1(b) shows the SHG wavelength shift in KTP crystals poled using Rb- and Ba-exchange. For Rb-exchange there is a 0.1 nm wavelength shift across the thickness of the crystal, which is not seen for the Ba-exchange. These results show that Ba-exchange is more suitable than Rb-exchange in experiments where the full aperture is needed, i.e., for high power frequency conversion, where a large diameter beam is required in the PPKTP. The technique may also be used in conjunction with, or complementary to Rb-exchange, for further development of sub-μm QPM structures for applications such as slow-light [5], ultra-bright biphotons [6], or squeezed-states generation [7].

References

[1] C. Liljestrand, F. Laurell, and C. Canalias, "Periodic poling of Rb-doped $KTiOPO_4$ by coercive field engineering", Opt. Express **24**, 14682-9 (2016).

[2] A. Zukauskas, A.-L. Viotti, C. Liljestrand, V. Pasiskevicius, and C. Canalias", Cascaded counter-propagating nonlinear interactions in highly-efficient sub-μm periodically poled crystals," Scientific Reports, **7**, 8037 (2017).

[3] P. Mutter, C. C. Kores, M. Widarsson, A. Zukauskas, F. Laurell, and C. Canalias, "Ion-exchanged waveguides in periodically poled Rb-doped $KTiOPO_4$ for efficient second harmonic generation", Opt. Express **28** (2020).

[4] K. Daneshvar, E. A. Giess, D. Kang, J. R. Williams, and D. Dawes, " Rb ion-exchange profile in potassium titanyl phosphate (KTP)", Opt. Mat. **12**, 453-457 (1999).

[5] J. B. Khurgin, "Slowing and stopping photons using backward frequency conversion in quasi-phase-matched waveguides," Phys. Rev. A **72**, 1–5 (2005).

[6] C. Chuu, and S. E. Harris, "Ultrabright backward-wave biphoton source," Phys. Rev. A **83**, 061803(R) (2011).

[7] L.-A. Wu, M. Xiao, and H. J. Kimble, "Squeezed states of light from an optical parametric oscillator", J. Opt. Soc. Am. B **4**, 1465 (1987).

Optical emission characterization of liquid core fibers filled with colloidal nanoplatelets

Simon Spelthann[1], Lars F. Klepzig[2,3], Dan Huy Chau[1], Mario Chemnitz[4,5], Saher Junaid[4], Ronja Stephan[1], Katharina Hausmann[1], Markus Schmidt[4,6], Jannika Lauth[2,3,7], Michael Steinke[1,8], Detlev Ristau[1,3,8]

1. Leibniz University Hannover, Institute of Quantum Optics, Welfengarten 1, D-30167 Hannover, Germany

2. Leibniz University Hannover, Institute of Physical Chemistry and Electrochemistry, Callinstraße 3A, D-30167 Hannover, Germany

3. Cluster of Excellence PhoenixD (Photonics, Optics, and Engineering – Innovation Across Disciplines), Welfengarten 1A, D-30167 Hannover, Germany

4. Leibniz Institute of Photonic Technology, Albert-Einstein-Straße 9, D-07745 Jena, Germany

5. INRS-EMT, 1650 Lionel-Boulet Blvd., Varennes, Quebec J3X 1P7, Canada

6. Otto Schott Institute of Material Research, Fraunhoferstraße 6, D-07745 Jena, Germany

7. Universität Tübingen, Institute of Physical and Theoretical Chemistry, Auf der Morgenstelle 18, D-72076 Tübingen, Germany

8. Leibniz University Hannover, QUEST-Leibniz-Research School, Callinstraße 36, D-30167 Hannover, Germany

Colloidal nanoplatelets have recently emerged as a novel and highly efficient emitters [1, 2]. Researchers especially exploited their remarkable optical properties such as large absorption and gain cross sections or narrow emission lines as solution-processed laser gain medium [3,4]. Recently, Maskoun and co-workers applied CdSe/CdS@$Cd_xZn_{1-x}S$ core/crown@gradient-alloyed shell colloidal semiconductor nanoplatelets dispersed in toluene as laser-active core in a microfluidic waveguide [4]. Here, we present an investigation of optical emission from a capillary fiber filled with solution-processed colloidal nanoplatelets. We synthesized two-dimensional colloidal CdSe/CdS core-crown nanoplatelets with a thickness of 4 and 5 monolayers and lateral dimensions of 10 – 60 nm by wet chemical methods, see Fig 1a) for a TEM picture. The nanoplatelets exhibit an emission feature at 515 and 550 nm, respectively, see Fig. 1b), which originates from their uniform thickness. In comparison to core-only nanoplatelets, the core-crown structure increases the photoluminescence quantum yield from <2% to 60 % (4ML) resp. from <1% to 45% (5ML) in toluene as is confirmed by steady-state

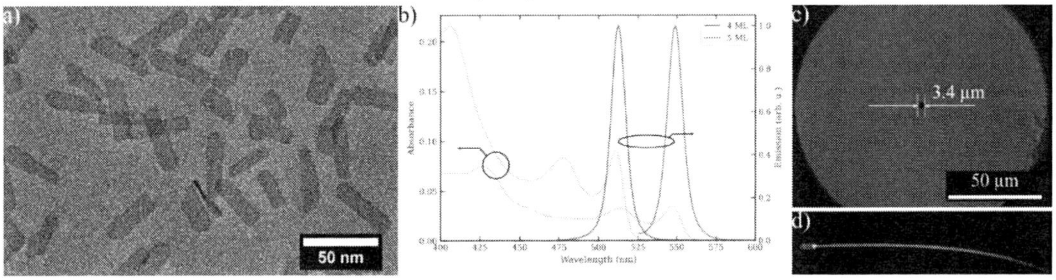

Fig 1 a) TEM image of the CdSe/CdS core-crown nanoplatelets, b) absorption and emission of the nanoplatelets with 4 and 5 monolayers, c) microsope image of the fiber cross section, d) filled fiber under excitation at 480 nm.

spectroscopy. We manufactured a capillary fiber by pressurizing a fused silica tube and drawing it to a fiber. The fiber has an outer diameter of 125 µm and an inner diameter of 3.4 µm, see Fig 1c). We mounted 30 cm fiber in optofluidic cells at both ends and injected the nanoplatelet dispersion in the reservoir of the cells. Capillary forces drew the dispersion into the fiber core and the filling level was verified under an optical microscope. We optically excited the nanoplatelets in the fiber using an optical parametric oscillator tuned to 480 nm. In our contribution, we will present an investigation of the emission characteristics and spectral behavior of the nanoplatelet-dispersion core fiber. The presented results will pave the way to developing the first nanoplatelet-filled fiber lasers.

Literature

[1] M. D. Tessier, P. Spinicelli, D. Dupoint, G. Patriarche, S. Ithurria, B. Dubertret, "Efficient Exciton Concentrators Built from Colloidal Core/Crown CdSe/CdS Semiconductor Nanoplatelets", Nano Lett. **14**, 207 (2014)
[2] F. Li, L. F. Klepzig, N. Keppler, P. Behrens, N. C. Bigall, H. Menzel, J. Lauth, "Layer-by-Layer Deposition of 2D CdSe/CdS Nanoplatelets and Polymers for Photoluminescent Composite Materials", *preprint, DOI:* 10.26434/chemrxiv-2022-m21qr (2022)
[3] B. Guzelturk, Y. Kelestemur, M. Olutas, S. Delikanli, H.V. Demir, "Amplified Spontaneous Emission and Lasing in Colloidal Nanoplatelets", ASC Nano **8**, 65599 (2014)
[4] C. She, I. Fedin, D.S. Dolzhnikov, P.D. Dahlberg, G.S. Engel, R.D. Schaller, D.V. Talapin, "Red, Yellow, Green, and Blue Amplified Spontaneous Emission and Lasing Using Colloidal CdSe Nanoplatelets", ASC Nano **9**, 9475 (2015)
[5] J. Maskoun, N. Gheshlaghi, F. Isik, S. Delikanli, O. Erdem, E.Y. Erdem, H.V. Demir, "Optical Microfluidic Waveguides and Solution Lasers of Colloidal Semiconductor Quantum Wells", Adv. Mater. **33**, 2007131 (2021)

Rapid THz-TDS Enabled by Single-Cavity Dual-Comb Gigahertz Laser

Benjamin Willenberg[1], Christopher R. Phillips[1], Justinas Pupeikis[1], Lars Liebermeister[2], Robert Kohlhaas[2], Björn Globisch[2], Ursula Keller[1]

1. Department of Physics, Institute for Quantumelectronics, ETH Zurich, Switzerland

2. Fraunhofer Institute for Telecommunications, Heinrich Hertz Institute, Berlin, Germany

Introduction: The THz spectral region is of high interest for scientific and industrial applications, enabling numerous non-destructive spectroscopy and thickness measurements. While dual comb lasers enable the optical delay sweeps needed for these measurements, most traditional systems are highly complex and have too long delay range. Here we resolve both these issues by demonstrating a single cavity dual-comb modelocked laser with GHz repetition rate based on a new multiplexing technique [1]. This new source supports nanosecond delay scans, sub-75-fs pulses, and multi-10-kilohertz update rates. We perform a proof of principle THz-TDS experiment by driving efficient photoconductive antenna devices [2] directly with the output of the laser oscillator.

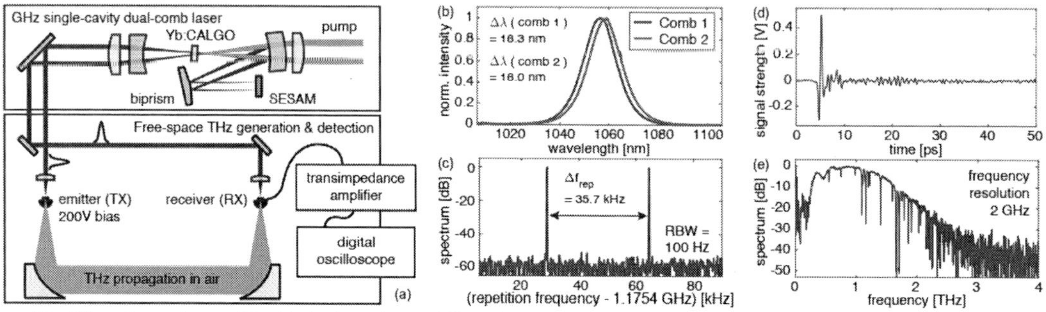

Fig. 1 Experimental setup (a) with single-cavity spatially multiplexed GHz dual-comb oscillator and free-space THz generation and detection. (b) Optical spectrum of the two combs. (c) Microwave spectrum of the two combs around the repetition rate of 1.175 GHz. (d) Averaged THz-signal time trace with a total integration time of 2 seconds. (e) Corresponding THz-spectrum with a frequency resolution of 2 GHz obtained from (d) via a 500-ps apodization window and Fourier transform.

Laser results: The dual-comb oscillator is shown in Fig. 1(a). Spatial multiplexing of the cavity is achieved with a biprism, yielding separate spots on the active elements (gain and SESAM). Additionally, the separation of the beams on the biprism allows continuous tuning of the repetition rate difference Δf_{rep} in a range larger than [-100, 100] kHz. The laser is pumped by a 980-nm diode delivering up to 900 mW of power from a single-mode fiber. The pump power is split equally and focused onto two spots in the Yb:CALGO gain crystal. Both combs show robust self-starting SESAM modelocking over a wide power range (from 60 mW to 100 mW average power per comb). They have a similar optical spectrum (Fig. 1 (b)) with a FWHM bandwidth >16 nm at a center wavelength of 1056 nm (comb 1) and 1058 nm (comb 2) and pulse duration below 75 fs. The fully shared cavity elements lead to common noise cancelation, and the high repetition rate enables a large Δf_{rep} without aliasing of the heterodyne beating between the two combs. Therefore, by using $|\Delta f_{rep}| > 20$ kHz, the two combs exhibit sufficient coherence in free-running operation to directly resolve the radio frequency comb lines and perform coherent dual-comb spectroscopy.

THz experiments: For the terahertz time domain measurements, we send the two combs onto free space THz antennas based on Fe-doped InGaAs [2]. After optical isolation and power control, up to 70 mW can reach each of the THz antenna. For the measurements we use 20 mW on the receiver and 70 mW on the electrically biased emitter. The THz radiation is collected and refocused with a pair of 2-inch parabolic mirrors yielding a path length in air of 265 mm. The receiver electro-optically samples the THz waveform via the delay sweep between the two combs. The generated current is amplified with a low-noise transimpedance amplifier and digitized on an oscilloscope. The THz frequencies are mapped to the RF domain via the dual-comb scaling factor $\Delta f_{rep}/f_{rep}$, where $\Delta f_{rep} = 35.7$ kHz. Using a trigger signal to define the time axis, the data from 71k periods (2 s integration time) is averaged in post processing to yield Fig. 1 (d). The distinct and sharp water absorption lines are visible in the spectrum, obtained by FFT from the time trace. The 500 ps apodization window restricts the spectral frequency resolution to 2 GHz. We achieve 40 dB dynamic range that allows to resolve features up to around 3 THz.

Conclusion: We have shown a single-cavity dual-comb gigahertz laser in free running operation with sub-75-fs pulse duration. The laser allows for direct observation of comb lines without processing on short time scale and with high Δf_{rep} avoiding any acoustic noise sources. We have applied the laser for a THz-TDS demonstration. Ultrafast ns scans, GHz resolution in THz domain, 40dB dynamic range with 2 s average.

References:

[1] J. Pupeikis, B. Willenberg, S. L. Camenzind, A. Benayad, P. Camy, C. R. Phillips, and U. Keller, "Spatially multiplexed single-cavity dual-comb laser for equivalent time sampling applications," arXiv, 2203.08536 (2022).

[2] B. Globisch, R. J. B. Dietz, R. B. Kohlhaas, T. Göbel, M. Schell, D. Alcer, M. Semtsiv, and W. T. Masselink, "Iron doped InGaAs: Competitive THz emitters and detectors fabricated from the same photoconductor", Journal of Applied Physics 121, 053102 (2017).

EPJ Web of Conferences **267**, 01039 (2022) https://doi.org/10.1051/epjconf/202226701039
EUROPHOTON 2022

Intra and extra-cavity beam shaping for post-compression of Yb:YAG picosecond high-energy pulses

Vincent Fortin[1], Marie-Christine Nadeau[1], Stéphane Petit[1]
1. CELIA Centre Lasers Intenses et Applications UMR5107, CNRS-Université de Bordeaux-CEA, 33400 Talence, France

Post-compression by self-phase modulation (SPM) enables to overcome the spectral bandwidth limitation of most laser gain materials enabling shorter powerful pulses. SPM in hollow-core fibers or multi-pass cells driven by high power lasers sources have shown impressive results [1,2]. However, post-compression of >100mJ pulses requires few meters-scale devices [3]. On the other hand, SPM in bulk thin-plates is a table-top sized solution to broaden the spectrum of high energy short pulses [4]. However, contrary to the multi-pass cells or hollow-core fiber where the post-compressed beam is spectrally homogeneous [5], the spatial homogeneity of the nonlinearities is a direct footprint of spatial distribution of the energy on the thin plate. Then, compared to gaussian beams, high energy flat-top beams directly shaped in the power amplifiers have shown highly homogeneous thin-plate post-compressed pulses [6]. Here we propose a new approach to perform thin plate pulse compression of high energy Yb systems with homogeneous spectrum by providing flat-top beams with the help of phase mirrors.

From this perspective, we investigated the flat-top beam generation from Ytterbium systems in two directions. Firstly, we explored the direct production of flat-top pulses in a Yb:YAG thin-disk cavity by developing a low-losses phase mirror. The cavity and the phase mirror have been designed to support the regenerative amplification of >50 mJ pulses at 1 kHz with a 4 mm-diameter flat-top spatial profile on the disk. Secondly, we investigated the extra-cavity conversion of gaussian beams to flat-top beams. In particularly, we will show experimental results of spectral broadening by SPM when we focused such shaped beam coming from a 1.6 mJ-400 fs Yb:KGW system at 1kHz on a silica plate. We will discuss about the spectral homogeneity obtained from this experiment.

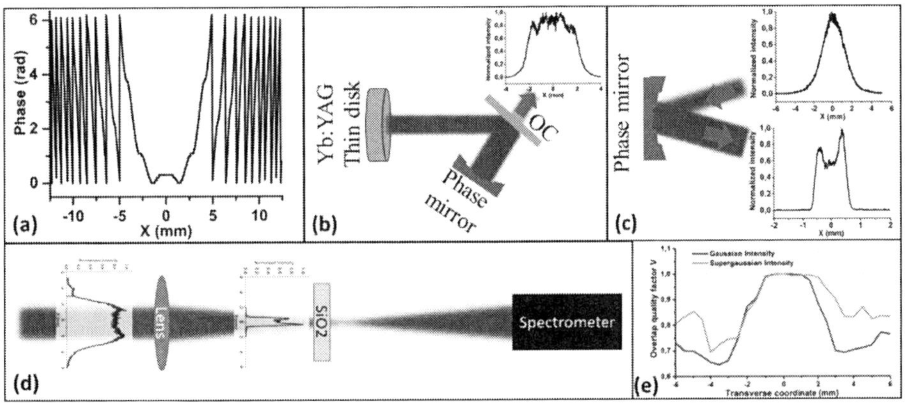

Fig. 1 : (a) Phase mirror phase designed for (b) intra-cavity and (c) extra cavity flat-top beam shaping. (d) Experimental setup for flat-top spectral broadening. (e) Gaussian versus supergaussian spectrum overlap quality factor

Finally, we will show how to apply this concept to the post-compression of higher energy pulses with HORIZON (1J-1kHz-1ps) currently under development. Starting from the Yb:YAG thin-disk regenerative amplifier front end delivering 50mJ pulses, we first designed a phase mirror to focus the compressed Gaussian beam in a flat-top beam at the intensity level of 1 TW/cm^2 on a 1-mm-thick silica plate. To post-compress the full 1J-pulses, we used another approach where we designed another phase mirror to shape the thin-disk regenerative amplifier beam in order to feed the final multipass power amplifier with an 8-mm-flat-top beam. In that case, the post-compression will be done just after the compressor without any further shaping. In both cases, the design of the phase mirror currently under manufacturing will be presented as well as the propagation and post-compression simulations. These results open the way to sub-100fs intense pulses from high energy powerful Yb:YAG systems.

References

[1] T. Nagy, P. Simon, et L. Veisz, « High-energy few-cycle pulses: post-compression techniques », *Adv. Phys. X*, vol. 6, n° 1, p. 1845795 (2021)
[2] A.-L. Viotti,M. Seidel, E. Escoto, S. Rajhans, W. P. Leemans, I. Hartl et C. M. Heyl, « Multi-pass cells for post-compression of ultrashort laser pulses », *Optica*, vol. 9, n° 2, p. 197-216 (2022)
[3] M. Kaumanns, D. Kormin, T. Nubbemeyer, V. Pervak, S. Karsch, et S. Karsch, « Spectral broadening of 112 mJ, 1.3 ps pulses at 5 kHz in a LG_{10} multipass cell with compressibility to 37 fs », *Opt. Lett.*, vol. 46, n° 5, p. 929-932 (2021)
[4] S. Y. Mironov, J. A. Wheeler, E. A. Khazanov, et G. A. Mourou, « Compression of high-power laser pulses using only multiple ultrathin plane plates », *Opt. Lett.*, vol. 46, n° 18, p. 4570-4573 (2021)
[5] J. Weitenberg, A. Vernaleken, J. Schulte, A. Ozawa, T. Satorius, V. Pervak, H.-D. Hoffman, T. Udem, P. Russbuldt, T. W. Hänsch, « Multi-pass-cell-based nonlinear pulse compression to 115 fs at 7.5 µJ pulse energy and 300 W average power », *Opt. Express*, vol. 25, n° 17, p. 20502-20510 (2017)
[6] S. Yu. Mironov,S. Fourmaux, P. Lassonde, V. N. Ginzburg, S. Payeur, J.-C. Kieffer, E. A. Khazanov, G. Mourou, « Thin plate compression of a sub-petawatt Ti:Sa laser pulses », *Appl. Phys. Lett.*, vol. 116, n° 24, p. 241101 (2020)

Spectroscopy and continuous wave laser operation of Tm^{3+}-doped $YScO_3$ mixed sesquioxide crystal

A. Suzuki[1,2,3], S. Kalusniak[3], H. Tanaka[3], M. Brützam[3], S. Ganschow[3], M. Tokurakawa[1,2], and C. Kränkel[3]

[1] *Institute for Laser Science, University of Electro-Communications, 1-5-1 Chofugaoka, Chofu Tokyo 182-8585, Japan*
[2] *Center for Neuroscience and Biomedical Engineering University of Electro-Communications, 1-5-1 Chofugaoka, Chofu Tokyo 182-8585, Japan*
[3] *Leibniz-Institut für Kristallzüchtung (IKZ), Max-Born-Straße 2, 12489 Berlin, Germany*

Tm^{3+}-doped mixed sesquioxides are attractive gain media for ultrafast 2-μm lasers due to their superior spectroscopic properties [1]. They exhibit a gain band at the wavelengths exceeding 2000 nm where no water vapor absorption and Tm^{3+} reabsorption is low, and their disordered crystal structure leads to broad and flat gain spectra due to inhomogeneous spectral broadening [2]. These features enable a very broad effective gain bandwidth, which is desirable for ultrashort pulse generation. In addition, high-quality large-size mixed sesquioxide crystals can be grown by the Czochralski method [3]. Conventional pure sesquioxide materials show excellent thermo-mechanical and spectroscopic properties as a laser medium [4], but they are difficult to fabricate due to their high melting points of more than 2350 °C. However, our current reinvestigation of the sesquioxide phase diagram revealed the existence of mixed sesquioxide compositions with melting temperatures below 2100 °C [3]. This enables the Czochralski growth from iridium crucibles, common for other oxide materials. Consequently, mixed sesquioxide crystals are promising candidates as commercially available gain materials for ultrafast lasers. Here, we present spectroscopic investigations and continuous wave laser operation of a Czochralski-grown Tm^{3+}-doped $YScO_3$ crystal.

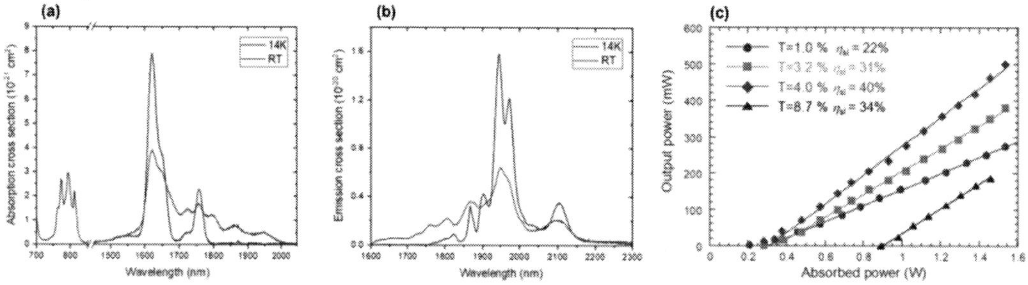

Fig. 1 (a) Absorption cross section and (b) Emission cross section at room temperature (RT) and 14 K.
(c) Output power characteristics as a function of the absorbed pump power with different OCs.

The laser crystal was prepared by the Czochralski growth from an iridium crucible from a melt composition of $(Tm_{0.02}Y_{0.5}Sc_{0.5})_2O_3$. The actual composition of the grown crystal was determined by inductively-coupled plasma optical emission spectroscopy to be $(Tm_{0.022}Y_{0.472}Sc_{0.506})_2O_3$. Figure 1 (a, b) show the absorption and emission cross sections at room temperature and 14 K. The absorption bands around 0.8 μm and 1.6 μm are useful for 2-for-1 pumping as well as in-band pumping for 2 μm laser operation, respectively. Broad and less-structured absorption and emission spectra are found even at low temperature, which indicates the expected inhomogeneous spectral broadening.

For the laser experiments, we constructed a 48 mm long plane-concave linear cavity. As a gain medium, a 6-mm thick sample was prepared, and a 780 nm single-emitter C-mount laser diode with a maximum output power of 3 W served as a pump source. Using output couplers (OCs) of different transmittance T, we obtained laser characteristics as shown in Fig. 1(c). Using a 4 % OC, the highest slope efficiency of 40% was achieved. The central wavelength was 2100 nm while the Stokes limit determined as the ratio of pump and laser wavelength amounts to only 37%. The highest slope efficiency of 40% was thus only possible by a cross relaxation process between adjacent Tm^{3+} ions in the 2-for-1 pumping scheme.

In conclusion, we investigated the spectroscopic properties and laser operation of a Tm^{3+}(2.2 at.%):$YScO_3$ mixed sesquioxide crystal. Its broadband absorption and emission spectra are favorable for ultrafast lasers, and continuous wave laser operation with a slope efficiency of 40% showed that laser-quality crystals can be grown by the Czochralski method.

References

[1] C. Kränkel et al., "Rare-earth doped mixed sesquioxides for ultrafast lasers," Opt. Mat. Express **12**, 1074-1091 (2022).
[2] R. Moncorgé et al., "Mid-infrared emission properties of the Tm^{3+}-doped sesquioxide crystals Y_2O_3, Lu_2O_3, Sc_2O_3 and mixed compounds $(Y,Lu,Sc)_2O_3$ around 1.5-, 2- and 2.3-μm," J. Lumin. **241**, 118537 (2022).
[3] C. Kränkel et al., "Czochralski growth of mixed cubic sesquioxide crystals in the ternary system Lu_2O_3–Sc_2O_3–Y_2O_3," Acta Cryst. **77**, 550-558 (2021).
[4] P. Loiko et al., "Highly efficient, compact Tm^{3+}:RE_2O_3 (RE = Y, Lu, Sc) sesquioxide lasers based on thermal guiding," IEEE J. Sel. Top. Quantum Electron. **24**, 1-13, (2018).

110 MW Thin-Disk Oscillator

Semyon Goncharov, Kilian Fritsch, Oleg Pronin
Helmut-Schmidt University, Holstenhofweg 85, 22043 Hamburg, Germany

Deep UV and XUV frequency combs can enable new applications in high precision spectroscopy and metrology. For instance, a low-lying isomer transition of thorium around 150 nm is an excellent candidate for a nuclear clock [1]. Additionally, the precise measurement of the 1s-2s transition in He^+ can facilitate important tests of quantum electrodynamics. These are just two applications to name. One promising approach to generate the required deep UV and XUV combs is high-harmonic generation driven by a femtosecond high power thin-disk laser-oscillator.

Here we present a driving thin-disk oscillator with 110 MW output peak power. Our compact, tabletop femtosecond Kerr-lens mode-locked thin-disk oscillator delivers the highest peak power among all mode-locked oscillators [2]. The system represents a diode-pumped Kerr-lens mode-locked Yb:YAG thin-disk laser (fig. 1a). The thin-disk is pumped by a 900 W, 940 nm fiber-coupled diode laser at a pump intensity of approximately 12 kW/cm^2. The resonator implements a telescope with two concave mirrors of -2500 mm radius of curvature, a 5 mm thick crystalline quartz plate placed under Brewster's angle close to the telescope's focus, and a hard aperture of 5 mm diameter (see fig. 1a). The laser is operated under low pressure to reduce the parasitic Kerr-lensing in the ambient air. 200 mbar of residual air pressure was the optimal working point to obtain an average output power of 202 W. The dispersion management is governed by two dispersively coated mirrors, introducing approximately -10000 fs^2 of total group delay dispersion (GDD) per round trip. Hence, it was possible to mode-lock the oscillator with 115 fs-long and 10 nm-wide (FWHM) bandwidth-limited pulses (fig 1b). The resonator operates at a repetition rate of 14 MHz resulting in 14.4 μJ pulse energy and, therefore 110 MW of extra-cavity peak power (0.6 GW intra-cavity). Single-pulsed operation is proven by a 1 m long (3.4 ns delay), home-built autocorrelator, and a fast photodiode (175-ps rise time, 2-GHz bandwidth) combined with a 2.5 GHz oscilloscope. The autocorrelation trace of the pulse was measured with a commercial ps-range autocorrelator (fig. 1b, inset). The beam quality factor M^2 was measured in accordance with ISO 11146 to be $M^2 = 1.1$. Additionally, the oscillator showed excellent average output power stability measured for 1 hour with 0.5 % RMS. A KLM oscillator with 102 MW peak power and 50 fs output pulse duration was recently demonstrated [3]. However, the oscillator can operate only for two minutes, making it impractical for applications.

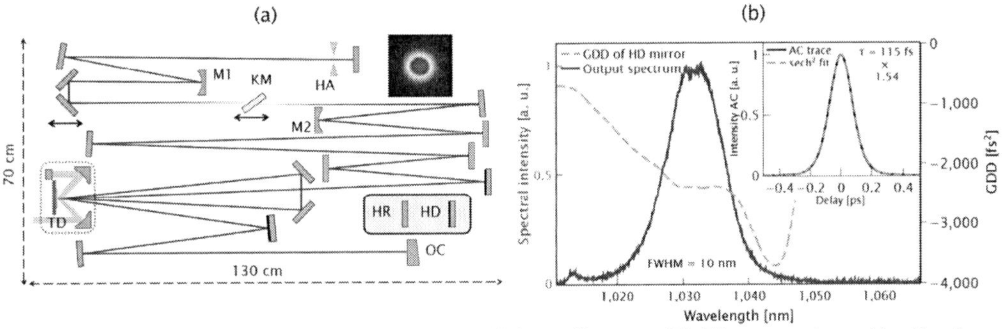

Fig. 1 (a) Schematics of the thin-disk oscillator: KM, 5 mm thick crystalline quartz; M1, M2, concave mirrors with radius of curvature -2500 mm; OC, 19 % transmission output coupler; HA, hard aperture; HR, high reflective mirrors; HD, highly-dispersive mirrors (-2400 fs^2 each); TD, thin-disk; **inset:** beam profile at 110 MW. **(b)** Spectrum of the output pulses with 110 MW peak power (red line) and GDD curve of HD mirrors (grey line), **inset:** intensity autocorrelation trace with sech^2-shaped fit.

This simple and compact high peak power thin-disk oscillator is the first step towards high repetition rate XUV frequency comb system. CEP stabilization of this oscillator in combination with highly efficient and robust spectral broadening and pulse compression in a multi-pass cell [4] are in progress. A GW-level peak power laser system delivering sub-20 fs pulses at 14 MHz rep. rate is in reach. Therefore, we believe our laser system will meet the highest demands of advanced spectroscopy in the XUV region in the near future.

References

[1] Seiferle, Benedict, et al. "Energy of the 229 Th nuclear clock transition." Nature **573**.7773: 243-246 (2019).
[2] Saraceno, Clara J., et al. "The amazing progress of high-power ultrafast thin-disk lasers." Journal of the European Optical Society-Rapid Publications **15**.1: 1-7 (2019).
[3] Fischer, Julian, et al. "Efficient 100-MW, 100-W, 50-fs-class Yb: YAG thin-disk laser oscillator." Optics Express **29**.25: 42075-42081 (2021).
[4] Weitenberg, Johannes, et al. "Nonlinear Pulse Compression to Sub-40 fs at $4.5~\mu $ J Pulse Energy by Multi-Pass-Cell Spectral Broadening." IEEE Journal of Quantum Electronics **53**.6, pp. 1-4, (2017).

50-W, >2-µJ SESAM-modelocked Ho:YAG thin-disk oscillator at 2.1 µm

Sergei Tomilov,[1,*] Yicheng Wang,[1] Martin Hoffmann,[1] Jonas Heidrich,[2] Matthias Golling,[2] Ursula Keller[2] And Clara J. Saraceno[1]

1. Photonics and Ultrafast Laser Science, Ruhr Universität Bochum, Universitätsstraße 150, 44801 Bochum, Germany

2. Department of Physics, Institute for Quantum Electronics, ETH Zürich, Auguste-Piccard-Hof 1, 8093 Zürich, Switzerland

**E-mail: sergei.tomilov@ruhr-uni-bochum.de*

Average power scaling of ultrafast laser systems operating in the short-wave infrared region (SWIR) from 1.4 µm to 3 µm continues to attract significant interest for a variety of scientific and industrial applications. In this spectral range, laser systems based on Ho-doped gain materials directly emitting at 2.1 µm are particularly attractive and have made significant progress, however so far only very few results were demonstrated in power-scalable laser geometries. In particular, the thin-disk geometry is promising for energy scaling of ultrafast amplifiers and for directly modelocked oscillators with high repetition rate and average power [1]. Previously, only one ultrafast TDL was demonstrated in this wavelength region based on Ho:YAG using Kerr-lens modelocking [2] where 25 W of average power and 0.325 µJ of pulse energy were achieved. More recently we showed record-high power single-mode CW operation up to 112 W [3]. Here, we demonstrate a SEmiconductor Saturable Absorber Mirror (SESAM) modelocked Ho:YAG thin-disk laser (TDL) emitting at 2.1-µm wavelength that achieves twice the average power and more than six time the pulse energy of previous demonstration, reaching 50 W of average power, with 1.13-ps pulses, 2.11 µJ of pulse energy and ~1.9 MW of peak power. To the best of our knowledge these are both the highest average power and pulse energy of any modelocked oscillator in this wavelength region to date.

The laser resonator used for modelocking is depicted in Fig. 1. It is based on a 2-at.%-doped and 190-µm-thick Ho:YAG disk pumped by a Tm-fiber laser with a maximum power of 209 W and emitting at 1908 nm. The laser beam passes the active medium twice in order to increase the available gain. For precise SPM control, we introduced a 2-mm thick undoped YAG plate on a translation stage and at Brewster's angle between the mirrors of a telescope, formed by two curved concave mirrors with a radius of curvature (RoC) of 0.3 m. All the flat folding intracavity mirrors are dispersive, introducing -21000 fs^2 of group delay dispersion per roundtrip. The high-power GaSb-based SESAM for this laser system was designed and grown using molecular beam epitaxy (MBE) at ETH Zurich [4].

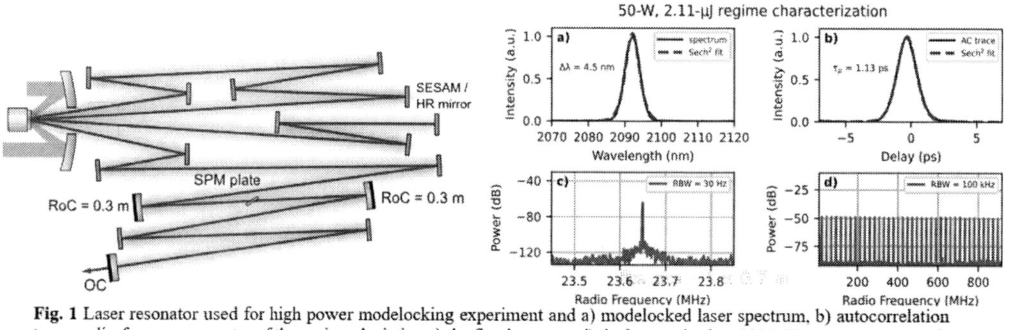

Fig. 1 Laser resonator used for high power modelocking experiment and a) modelocked laser spectrum, b) autocorrelation trace, radio-frequency spectra of the regime depicting c) the first beatnote, d) the harmonics in a 900 MHz span. RBW stands for resolution bandwidth.

At the full pump power using the 2% output coupler we achieved stable soliton modelocking at an average power of 50 W at a repetition rate of ~23.7 MHz. At this power level, the pulses had a duration of 1.13 ps assuming sech^2 pulse shape, with 4.5-nm spectral bandwidth, corresponding to a time-bandwidth product of 0.348. The radio-frequency spectrum confirms stable modelocking with no modulations within the measurement range. Single-pulse operation was verified using a fast photo-detector and sampling oscilloscope.

We are confident that reducing the pulse duration to the few hundred femtoseconds and increasing the available pulse energy of this system towards the 10-µJ level in future should be straightforward with tailored SESAM designs. In its current implementation, external pulse compression can already bring this system to scientific applications.

References

[1] C. J. Saraceno, D. Sutter, T. Metzger, and M. Abdou Ahmed, "The amazing progress of high-power ultrafast thin-disk lasers," J. Eur. Opt. Soc.-Rapid Publ. **15**(1), 15 (2019);

[2] J. Zhang, K. F. Mak, and O. Pronin, "Kerr-Lens Mode-Locked 2-µm Thin-Disk Lasers," IEEE J. Sel. Top. Quantum Electron. **24**(5), 1-11 (2018);

[3] S. Tomilov, M. Hoffmann, Y. Wang, and C. Saraceno, "Moving towards high-power thin-disk lasers in the 2-µm wavelength range," J. Phys. Photonics **3**, 022002 (2021);

[4] J. Heidrich, M. Gaulke, B. O. Alaydin, M. Golling, A. Barh, and U. Keller, "Full optical SESAM characterization methods in the 1.9 to 3-µm wavelength regime," Opt. Express **29**(5), 6647 (2021).

Third Harmonic Generation and $\chi^{(5)}$ Effects in Thin Gradient HfO_2 Layers

David Zuber[1,2], Sven Kleinert[1,2], Ayhan Tajalli[1,2,3], Morten Steinecke[4], Marco Jupé[2,4], Ihar Babushkin[1,2,5], Detlev Ristau[1,2,4], Uwe Morgner[1,2,4]

1. Leibniz Universität Hannover, Welfengarten 1, 30167 Hannover, Germany
2. Cluster of Excellence PhoenixD (Photonics, Optics, and Engineering-Innovation Across Disciplines), 30167, Hannover, Germany
3. Currently in Deutsches Elektronen-Synchrotron DESY, Notkestraße 85, 22607 Hamburg, Germany
4. Laser Zentrum Hannover e.V., Hollerithallee 8, 30419 Hannover, Germany
5. Max Born Institute, Max-Born-Strasse 2a, 10117 Berlin, Germany

Nowadays, frequency tripling of near infrared lasers is a typical way to generate high power, ultrashort UV pulses [1]. However, third harmonic generation (THG) suffers either from a low efficiency if the TH is driven directly in bulk crystals [2], or from a complicated and alignment sensitive setup if cascaded second order effects are used. Instead of these approaches, third harmonic generation in thin-layered structures feature a much smaller footprint and relaxed phase matching condition [3]. There are also clear indications that their biggest drawback, the low conversion efficiency typically caused by the small interaction length in the layers, will be overcome in near future. Recently, a conversion efficiency of 1.8% had been achieved for a multilayer system with Hafnia as nonlinear material component [4]. Besides a careful optimization of the layer structure, also the chosen materials and their nonlinear properties play an important role. In this work, we present theoretical and experimental studies of the nonlinear properties of ion beam sputtered HfO_2-layers. To get a better insight into the THG effects, the material is deposited as a gradient layer with continuously varying thickness from 200 nm to 1500 nm. A home built femtosecond laser system at 1030 nm [5] is focused onto the sample (see Fig. 1(a)), and the TH is recorded as a function of the layer thickness. In parallel, FDTD simulations based on the software 'Lumerical' are performed, including linear and third order nonlinear effects of the layer and the substrate material. The results are shown in Fig. 1(b). Simulation and experiment show good agreement and correctly reflect the interference effects caused by both, the fundamental and the third harmonic wave, as well as linear and nonlinear influences caused by the substrate.

The experiments were performed at intensities in the range of several hundred GW/cm^2. By increasing the intensity, the TH signal increases stronger than expected for a third-order effect. While this effect can partially be attributed to self-focussing in the substrate, also further effects seems to have an influence on the THG. After careful consideration of different linear and nonlinear effects, THG by fifth order nonlinearity ($3\omega = \omega + \omega + \omega + \omega - \omega$) seems to be the most likely explanation. By comparison of the experimentally measured TH to the simulations, values of $\chi^{(3)} = (1.0 \pm 0.17) \times 10^{-21}\, m^2/V^2$ and $\chi^{(5)} = (3.1 \pm 1.9) \times 10^{-40}\, m^4/V^4$ are obtained. Up to the authors knowledge, this is the first time a value of the fifth order nonlinear susceptibility is derived for dielectric coating materials. We believe that this detailed study of involved effects on the THG in thin layers will lead to a new generation of multi-layered structures for THG with even higher conversion efficiencies.

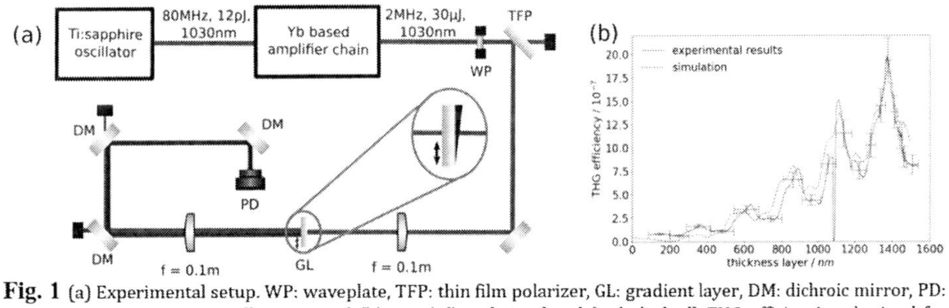

Fig. 1 (a) Experimental setup. WP: waveplate, TFP: thin film polarizer, GL: gradient layer, DM: dichroic mirror, PD: photodiode. (b) Experimentally measured (blue, solid) and simulated (red, dashed) THG efficiencies obtained for different layer thickness under consideration of linear and nonlinear effects in both, the layers material and the substrate.

References

[1] J. Rothhardt et al., "100 W average power femtosecond laser at 343 nm," Opt. Lett. **41**, 1885 (2016).
[2] K. Miyata et al., "High efficiency third harmonic generation in BiB_3O_6," Opt. Lett. **36**, 3627 (2011).
[3] C. Rodríguez et al., "Frequency tripling mirror," Opt. Express **23**, 31594 (2015).
[4] S. Balendat et al., "Manufacturing and characterization of frequency tripling mirrors," in 2021 Conference on Lasers and Electro-Optics Europe and European Quantum Electronics Conference, OSA Technical Digest (Optica Publishing Group, 2021), paper cd_p_33.
[5] T. Binhammer et al., "CPA-free high repetition rate few-cycle OPCPA," in 2019 Conference on Lasers and Electro-Optics Europe and European Quantum Electronics Conference, OSA Technical Digest (Optica Publishing Group, 2019), paper cg_p_20.

https://doi.org/10.1051/epjconf/202226701044

Direct broadband infrared generation from 12 to 35 THz with a Kerr-lens modelocked Cr:ZnS oscillator

Johann Gabriel Meyer[1], Oleg Pronin[1]

1. Helmut-Schmidt-Universität, Holstenhofweg 85, 22043 Hamburg, Germany

Broadband mid-infrared sources are demanded tools for spectroscopy in the molecular fingerprint region, potentially enabling detection of biomarkers for medical diseases via blood and breath analysis [1]. Spectroscopic methods like nano-FTIR or electro-optic sampling offer here broadband spectral coverage together with unique spatial resolution or high sensitivity and ultrafast temporal resolution [2,3]. These methods require high brightness and broadband infrared sources. Broadband pulses from Cr:ZnS lasers are an excellent choice for the relatively simple and robust process of intra-pulse difference frequency generation (IDFG) [4]. We present here mid-IR generation with a spectral coverage from 12 to 35 THz (9 – 25 μm) based on IDFG in a GaSe crystal with the pulses directly coming from a Kerr-lens modelocked (KLM) Cr:ZnS oscillator.

The developed KLM Cr:ZnS oscillator delivers pulses with a wavelength range from 2.1 to 2.5 μm (-10 dB), corresponding to a bandwidth of 23 THz. The spectrum supports a pulse duration of 37 fs (FWHM). FROG measurements of a similar oscillator configuration have confirmed pulse durations down to 36 fs. The oscillator runs at a repetition rate of 70 MHz and delivers an average power of 1.55 W, resulting in a pulse energy of 22 nJ and a peak power of 527 kW. The oscillator operates in negative dispersion regime with overall GDD and TOD compensated by dispersive mirrors. The OC mirror transmits 35 %. It is coated on a low dispersive CaF_2 substrate.

Mid-IR generation was already observable when the oscillator pulses are directly focused (f = 25 mm) into the nonlinear GaSe crystal. At wavelengths around 2 μm, GaSe exhibits an advantage of an increased damage threshold and a broader phase-matching bandwidth when compared to the standard wavelengths around 1 μm. Furthermore, GaSe shows transparency up to a wavelength of 20 μm. When employing a ca. 0.5 mm thick GaSe crystal, broadband mid-IR radiation was generated covering the range from 12 to 35 THz (9 – 25 μm) [s. Fig. 1 (a)]. The total mid-IR power lies in the range of several μW. When radiation from a similar oscillator with an average power of 1 W was amplified by a factor two and spectrally broadened in a simple single pass amplifier based on another 9 mm long Cr:ZnS crystal, the mid-IR power could be extended above 0.2 mW.

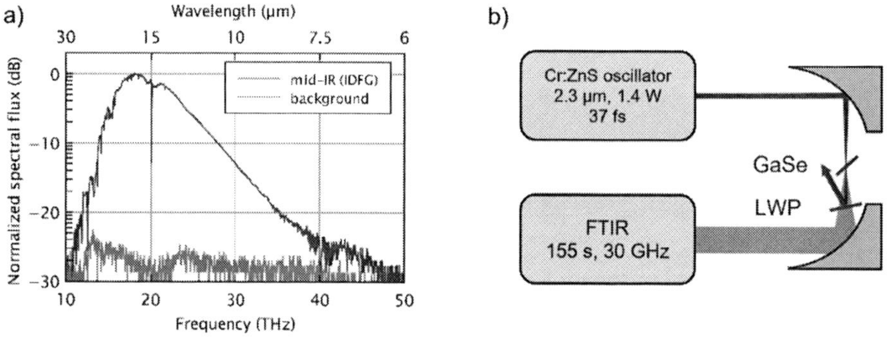

Figure 1. (a) FTIR-spectrum of the mid-IR radiation generated via intra-pulse difference frequency generation (IDFG) in GaSe directly from pulses of a Kerr-lens modelocked Cr:ZnS oscillator (red curve). The prominent absorption line of atmospheric CO_2 is appearing at 20 THz (15 μm) together with the weaker, neighboring R- and P-branches. The background was recorded after disturbing the modelocking operation (gray curve). The spectra were averaged from 50 scans recorded over a duration of 155 s. The spectral resolution of the measurements is 30 GHz. (b) Setup for direct mid-IR generation. The oscillator radiation was focused by an OAP mirror with a focal length of 25 mm into a ca. 0.5 thick GaSe crystal. Afterwards, it is separated by a longwave pass (LWP) filter with a cut-off at 3.6 μm. The generated mid-IR radiation was recorded with an FTIR spectrometer.

In summary, we have developed a KLM Cr:ZnS oscillator and generated directly broadband, coherent mid-IR radiation via IDFG in GaSe ranging from 12 to 35 THz (-20 dB). Due to the IDFG process, the radiation has an inherently stable carrier-envelope offset frequency and represents an excellent source for further spectroscopic studies based on electro-optic sampling.

References

[1] Huber, M., Kepesidis, K.V., Voronina, L. *et al.* "Stability of person-specific blood-based infrared molecular fingerprints opens up prospects for health monitoring," *Nat. Commun.* **12,** 1511 (2021).
[2] Peller, D., Roelcke, C., Kastner, L.Z. *et al.* „Quantitative sampling of atomic-scale electromagnetic waveforms," *Nat. Photonics* **15,** 143–147 (2021).
[3] Steinleitner, P., Nagl, N., Kowalczyk, M. *et al.* "Single-cycle infrared waveform control," *Nat. Photon.* (2022).
[4] Qing Wang, Jinwei Zhang, Alexander Kessel, Nathalie Nagl, Vladimir Pervak, Oleg Pronin, and Ka Fai Mak, "Broadband mid-infrared coverage (2–17 μm) with few-cycle pulses via cascaded parametric processes," Opt. Lett. **44,** 2566-2569 (2019).

 https://doi.org/10.1051/epjconf/202226701045

Inline Amplification of Mid-Infrared Intrapulse Difference Frequency Generation

Quentin Bournet[1,2], Florent Guichard[2], Michele Natile[2], Yoann Zaouter[2], Antoine Zheng[1], Manuel Joffre[3], Adeline Bonvalet[3], Mindaugas Jonusas[3], Frédéric Druon[1], Marc Hanna[1], and Patrick Georges[1]

1. *Université Paris-Saclay, Institut d'Optique Graduate School, CNRS, Laboratoire Charles Fabry, 91127 Palaiseau, France*
2. *Amplitude, 11 Avenue de Canteranne, Cité de la Photonique, 33600 Pessac, France*
3. *Laboratoire d'Optique et Biosciences, Ecole Polytechnique, CNRS, INSERM, Institut Polytechnique de Paris, 91128 Palaiseau, France*

We demonstrate an all inline two stages mid-infrared (MIR) source based on intrapulse difference frequency generation (iDFG) and subsequent optical parametric amplification (OPA), with pump recycling. The source is driven by a nonlinearly compressed high-energy Yb-doped-fiber amplifier delivering 7.4 fs pulses at a central wavelength of 1030 nm, at a repetition rate of 250 kHz [1]. It delivers 1 μJ 73 fs (sub-three optical cycles) pulses at a central wavelength of 8 μm, corresponding to a total efficiency of 2 %. The amplified MIR central wavelength is tunable over more than one octave ranging between 4-8 μm.

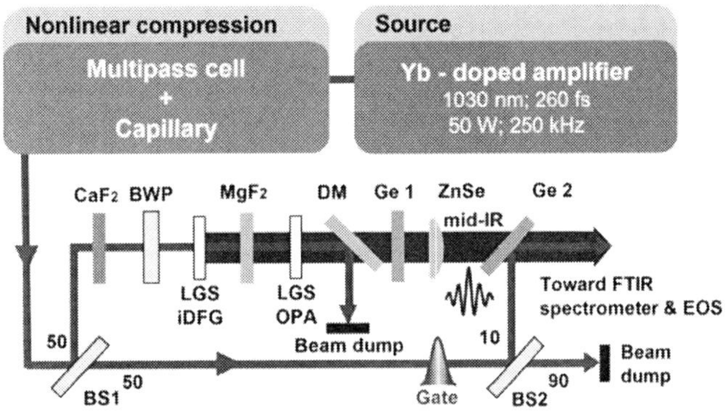

Fig. 1 Schematic of the experimental setup. The pump beam is sent on the 50/50 beamsplitter (BS1). The reflected driving pulses go through calcium fluoride (CaF2) windows to adjust the chirp and the bichromatic waveplate (BWP). Then, they drive the generation in the XY LGS crystal and the amplification in the XZ LGS crystal. The MIR radiation is filtered by a dichroic mirror (DM). A germanium plate (Ge 1) is inserted to adjust the GDD of the MIR and a zinc selenide (ZnSe) lens is used to collimate it. A magnesium fluoride (MgF2) crystal is inserted to adjust the delay. The transmitted pulses (gate) are sent to a second beamsplitter (BS2) and to a second germanium plate (Ge 2) to recombine the gate and the MIR for EOS measurement.

At the iDFG stage output, the 1030 nm residual driving pump pulses are recycled to transfer part of its energy to amplify the MIR pulses in the OPA stage. As shown in Fig. 1, this ultrafast MIR source combines a chirp-management calcium fluoride crystal, a bichromatic waveplate (BWP) [1,2], a type II $LiGaS_2$ (LGS) crystal-based iDFG stage, a delay-management magnesium fluoride birefringent crystal, and a type I LGS crystal-based OPA stage. The high efficiency of the source over the whole MIR central wavelength tunability range is ensured by a precise control of the different nonlinear processes involved in each stage. The setup allows an independent management of the different wavelength delays to obtain perfect interaction pulses overlap in the OPA step.

This implementation results in several key features: (i) all pump photons are polarized adequately to be enrolled in the iDFG process (ii) the long wavelength pump photons that have been amplified in the iDFG stage are recycled and used to pump the OPA stage after a resynchronization stage (iii) the architecture is all inline, with no spatial and temporal overlap alignment involved. This architecture is compact and simple, combining high energy, broad bandwidth, and high repetition rate, particularly suitable for nonlinear time-resolved 2D spectroscopy experiments and drive High Harmonic Generation (HHG) in solids.

References

[1] Q. Bournet, F. Guichard, M. Natile, Y. Zaouter, M. Joffre, A. Bonvalet, I. Pupeza, C. Hofer, F. Druon, M. Hanna et al., "Enhanced intrapulse difference frequency generation in the mid-infrared by a spectrally dependent polarization state," Opt. Lett. **47**, 261–264 (2022).
[2] H. Fattahi, A. Schwarz, S. Keiber, and N. Karpowicz, "Efficient, octave-spanning difference-frequency generation using few-cycle pulses in simple collinear geometry," Opt. letters **38**, 4216–4219 (2013).

1 kHz Yb:YAG thin-disk high-energy picosecond regenerative amplifier

Marie-Christine Nadeau[1], Philippe Balcou[1], Dominique Descamps[1], Christophe Féral[1], Vincent Fortin[1], Jérôme Lhermite[1], Denis Marion[1], Éric Mével[1], Antoine Rohm[1], Stéphane Petit[1]

1. CELIA Centre Lasers Intenses et Applications UMR5107, CNRS-Université de Bordeaux-CEA, 33400 Talence, France

1J-1kW-1ps is a real scientific and technological challenge involving thermal, energy and short pulse management. Yb:YAG is nowadays the best-suited amplifier material to address these challenges. Recently, 1.1 J at 1 kHz has been reported with cryogenic-cooled thick Yb:YAG disks but at a longer pulse duration of 4.5 ps [1]. On the other hand, shorter pulses of 920 fs have been obtained by the solely thin-disk technology with an energy of 720 mJ at 1 kHz [2]. In the frame of laser development (HORIZON) for new high energy secondary sources and plasma physics, CELIA has chosen an alternative way combining thin-disk technology to provide a high energy front-end and rotating water-cooled disk technology for the final power amplifier. Although the best performances of Yb:YAG thin-disk regenerative amplifiers reach 200 mJ at 1-5 kHz [2,3], most of the published schemes are rather complex and may leave key issues unreported. Here, we report on the development of the Yb:YAG thin-disk front-end based on a simple scheme with the experimental results and the faced issues.

The regenerative cavity is based on a 220-µm 7%-doped Yb:YAG thin-disk CW diode-pumped at 969 nm. The 12 mm disk is mounted on a diamond heat-sink and it has a 4 m radius of curvature when not pumped. The thin-disk is placed at one end of a 4.8 m long linear cavity. A magnification telescope is inserted in such a way that the fundamental mode has a 4.8 mm diameter on the disk matching 86% of the flat-top pump diameter. The regenerative operation is insured by a thin-film polarizer combined with a large aperture BBO single crystal ($12x12x40\ mm^3$) operating at 1 kHz. The cavity is seeded by 1.2 mJ-1.4 ns stretched pulses with a 2.5 nm spectral bandwidth (FWHM).

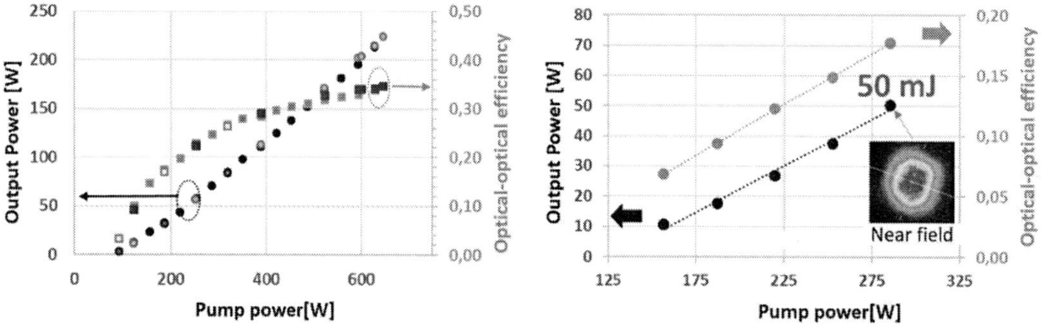

Fig. 1 Output power and efficiency as a function of the pump power in (a) fundamental mode CW operation and (b) seeded regenerative operation at 1 kHz.

In CW and fundamental mode operation, the cavity delivers more than 220 W of output power with 35% optical-optical efficiency with a beam quality factor M^2=1.08. In regenerative operation, an amplified pulse energy up to 50 mJ has been obtained with a non- saturated optical-optical efficiency of 18% which is, to the best of our knowledge, the best performance ever reported for a CW-pumped regenerative cavity with a single Yb:YAG thin-disk head. The amplified pulses have been then recompressed down to 900 fs. Finally, we will show also how the thermal issues occurring in the Pockels-cell limit the output power of the regenerative amplifier. Despite an extremely low single-pass absorption of the BBO crystal, its temperature increases beyond 40 °C at 50 W output power degrading the long-term voltage contrast ratio of the Pockel's cell, the beam quality and the output power stability. To the best of our knowledge, such behaviour in regenerative amplifiers has never been reported and we will discuss how to address this issue to further increase the output energy.

References

[1] Y. Wang, H. Chi, C. Baumgarten, K. Dehne, A. R. Meadows, A. Davenport, G. Murray, B. A. Reagan, C. S. Menoni, and J. J. Rocca "1.1 J Yb:YAG picosecond laser at 1 kHz repetition rate, " Opt. Lett., vol. 45, pp. 6615-6618, 2020.

[2] C. Herkommer, P. Krötz, R. Jung, S. Klingebiel, C. Wandt, R. Bessing, P. Walch, T. Produit, K. Michel, D. Bauer, R. Kienberger, and T. Metzger, "Ultrafast thin-disk multipass amplifier with 720 mJ operating at kilohertz repetition rate for applications in atmospheric research," Opt. Exp., vol. 28, pp. 30164-30173, 2020.

[3] T. Nubbemeyer, M. Kaumanns, M. Ueffing, M. Gorjan, A. Alismail, H. Fattahi, J. Brons, O. Pronin, H. G. Barros, Z. Major, T. Metzger, D. Sutter, and F. Krausz,, "1 kW, 200 mJ picosecond thin-disk laser system," Opt. Lett., vol. 42, pp. 1381-1384, 2017

https://doi.org/10.1051/epjconf/202226701047

Optimized composition of $LiREF_4$ (RE = Tb_xY_{1-x}) crystals for efficient green and yellow lasers - fluorescence quenching in Tb^{3+} ions

M. Badtke, S. Kalusniak, S. Püschel, H. Tanaka, C. Kränkel
Leibniz-Institut für Kristallzüchtung (IKZ), Max-Born-Straße 2, 12489 Berlin, Germany

Achieving visible laser emission is often challenging, in particular in the yellow spectral range. The range between 550 and 600 nm is not covered by semiconductor laser sources. Dye laser systems are complex and inconvenient to use. As a result, coherent yellow radiation is mainly achieved by nonlinear frequency conversion of near-infrared lasers, but the efficiencies into the yellow are typically low. In recent years solid-state lasers based on trivalent Tb^{3+} have been shown to be a promising alternative for direct green and yellow lasers [1]. In typical fluoride host crystals, like $LiYF_4$ or $LiLuF_4$, pump-limited output powers of more than 1 W and 0.5 W were realized in the green and yellow, respectively. However, the slope efficiencies around 60% and 25% are far below the theoretical limits of 90% and 83%, respectively, for reasons not understood in detail, yet [2].

A possible explanation is a reduced quantum efficiency of the emitting $^5D_4$ level by an inversion dependent loss process. To investigate this, we present a detailed study on the fluorescence quenching of Tb^{3+} in the $LiYF_4$ host matrix for different doping concentrations up to stochiometric $LiTbF_4$ by adopting a Z-scan measurement technique. Figure 1 shows decay curves of fluorescence for a stoichiometric $LiTbF_4$ and a $LiTb_{0.28}Y_{0.72}F_4$ for different pump intensities at a wavelength of 488 nm. At low pump intensities both crystals exhibited single exponential decays with a fluorescence lifetime of 4.8 and 4.9 ms, respectively.

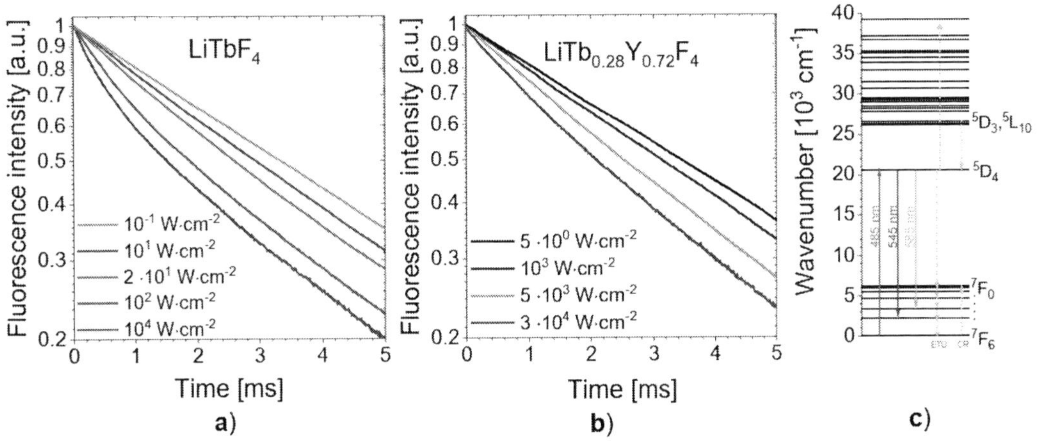

Fig. 1 Temporal decay of the fluorescence from the $^5D_4$ energy level multiplet, in a logarithmic scale, recorded at different pump intensities at a wavelength of 488 nm. **a)** $LiTbF_4$, **b)** $LiTb_{0.28}Y_{0.72}F_4$, **c)** Energy diagram of Tb^{3+} in $LiYF_4$.

At high pump intensities a quenching of the lifetime occurs, indeed revealing an inversion dependent quenching process, which we attribute to energy transfer upconversion (ETU). We also observed the onset of fluorescence quenching at different densities of excited Tb^{3+} ions in samples with different Tb^{3+} doping concentrations. Such behavior can be only understood when assuming the upconversion process to be mediated by energy migration between Tb^{3+} ions, a process depending on the doping concentrations. To describe such migration-assisted upconversion process we adopted a model for nonlinear luminescence quenching [3], modified for a fast relaxation of the upconverted ion [4]. Least square fitting with this model yields an excellent agreement with our data, resulting in values for donor-donor interaction and donor acceptor interaction parameters, which describe the strength of the ETU and energy migration between Tb^{3+} ions, respectively. From these parameters, we estimate the quantum efficiency with respect to the doping concentration and inversion level. Ultimately, this provides a guideline for the ideal doping concentration of Tb^{3+} for efficient laser operation by suppressing of the quenching process.

References

[1] P. W. Metz et al., "Efficient continuous wave laser operation of Tb^{3+}-doped fluoride crystals in the green and yellow spectral regions", Laser Photon. Rev. **10**, 335-344 (2016).
[2] S. Kalusniak et al., "Spectroscopic properties of Tb^{3+} as an ion for visible lasers". Appl. Phys. B **128**, 33 (2022).
[3] D. A. Zubenko et al., "Different mechanisms of nonlinear quenching of luminescence ", Phys. Rev. B **55**, 8881 (1997).
[4] L. Agazzi et al., "Energy-transfer-upconversion models, their applicability and breakdown in the presence of spectroscopically distinct ion classes: A case study in amorphous Al_2O_3:Er^{3+}", J. Phys. Chem. C **117**, 6759-6776 (2013).

https://doi.org/10.1051/epjconf/202226701048

Generation and control of single-cycle mid-infrared waveforms

Maciej Kowalczyk[1,2,3], Philipp Steinleitner[1], Nathalie Nagl[1,2], Jinwei Zhang[1,4], Vladimir Pervak[2], Christina Hofer[1,2,3], Aleksander Głuszek[5], Jarosław Sotor[5], Ferenc Krausz[1,2,3], Alexander Weigel[1,3,*], Ka Fai Mak[1]

1. *Max-Planck-Institut für Quantenoptik, Hans-Kopfermann-Str. 1, 85748 Garching, Germany*
2. *Fakultät für Physik, Ludwig-Maximilians-Universität München, Am Coulombwall 1, 85748 Garching, Germany*
3. *Center for Molecular Fingerprinting, 1093 Budapest, Czuczor utca 2-10, Hungary*
4. *School of Optical and Electronic Information & Wuhan National Laboratory for Optoelectronics, Huazhong University of Science and Technology, Wuhan, China, Luoyu Road 1037, 430074, Wuhan, China*
5. *Laser & Fiber Electronics Group, Faculty of Electronics, Photonics and Microsystems, Wroclaw University of Science and Technology, Wybrzeże Wyspiańskiego 27, 50-370 Wrocław, Poland*

**E-mail: alexander.weigel@mpq.mpg.de*

The generation of ultrashort laser pulses comprising a mere few or even a single electric-field oscillation, and the possibility to shape such electric-field waveforms are key technologies for investigating and controlling fundamental light-matter interactions on the sub-femtosecond time scale [1]. Extending the sub-cycle waveform control from the previously-demonstrated near-infrared (near-IR) to the mid-infrared (mid-IR) spectral range is crucial for several applications, including next-generation light-wave-driven electronic signal processing in low-bandgap highly-doped semiconductors [2] and field-resolved spectroscopy of biological systems [3].

Here, we present a novel approach for the generation and control of single-cycle mid-IR waveforms. Down-conversion of carrier-envelope-phase (CEP) stable single-cycle pulses from a Cr:ZnS laser (Fig. 1a) in a single $ZnGeP_2$ (ZGP) crystal provides a 3.7-octave supercontinuum, stretching from the near-IR to the mid-IR fingerprint region (Fig. 1b). By exploiting the CEP-dependence of the underlying cascaded intra-pulse difference-frequency-generation (IPDFG) we can continuously adjust the waveforms of the single-cycle mid-IR transients (Fig. 1c). The unprecedented CEP-stability of our Cr:ZnS driving laser results in highly reproducible control over the waveform generation [4].

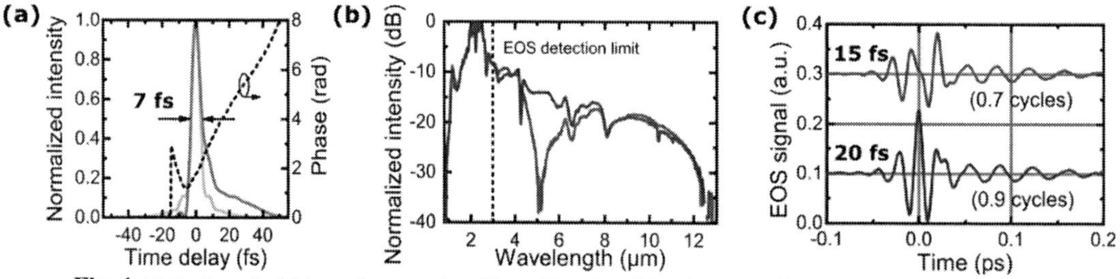

Fig. 1 (a) Single-cycle driving pulse as retrieved from SHG-FROG (green) compared with the Fourier-limited pulse (grey). (b) Supercontinuum from cascaded IPDFG for the extreme cases of CEP-dependent constructive (blue) and destructive (red) interference between the even and odd cascaded IPDFG orders. EOS measurement with 7-fs gating pulse allows to measure components with wavelengths longer than ~3μm. (c) EOS traces of the supercontinuum pulses in *b*; the FWHM pulse durations and the number of optical cycles at the spectral centroid are indicated.

The frontend of our setup is a diode-pumped Kerr-lens mode-locked Cr:ZnS laser oscillator producing 28 fs pulses at 2.3 μm. The repetition rate and the average output power amounted to 23 MHz and 480 mW, respectively. The laser output was focused into a 0.5-mm-thick TiO_2 plate, providing spectral broadening to a super-octave-spanning bandwidth (1.1 – 3.2 μm), with a corresponding Fourier-limit of 6 fs. The pulses were compressed with custom-designed multilayer dispersive mirrors down to 7 fs (Fig. 1a), constituting the first demonstration of single-cycle pulses generated with chromium-doped II-VI laser technology. These single-cycle pulses were CEP-stabilized with a residual phase noise of only 11 mrad (integrated from 1 mHz to 11.5 MHz). Subsequently, the beam was focused into a ZGP nonlinear crystal for cascaded IPDFG, resulting in an infrared supercontinnum covering 3.7 optical octaves (0.9 – 12 μm; Fig. 1b). The output power of the mid-IR ($\lambda > 3.6$ μm) amounted to 31 mW, corresponding to a high IPDFG conversion efficiency of 14%. We characterized the mid-IR transients with electro-optical sampling, using the 7-fs driver pulses for gating. Finally, we demonstrate that by tuning the CEP of the stabilized Cr:ZnS driver laser we can continuously shape the single-cycle mid-IR waveforms (Fig. 1c) and modify their spectral intensity (Fig. 1b).

References

[1] T. Brabec and F. Krausz, "Intense few-cycle laser fields: Frontiers of nonlinear optics," Rev. Mod. Phys. **72**, 545–591 (2000).
[2] M. P. Fischer et al., "Field-resolved detection of the temporal response of a single plasmonic antenna in the mid-infrared," Optica **8**, 898-903 (2021).
[3] I. Pupeza, M. Huber et al., "Field-resolved infrared spectroscopy of biological systems," Nature **577**, 52–59 (2020).
[4] P. Steinleitner, N. Nagl, M. Kowalczyk et al., "Single-cycle infrared waveform control", Nat. Photon. (2022).

Entirely passive thin-disk dual-comb spectrometer operating in green

Tobias Hofer[1], Kilian Fritsch[1] and Oleg Pronin[1]

1. Helmut-Schmidt University, Holstenhofweg 85, 22043 Hamburg, Germany

Dual-comb spectroscopy (DCS) is an attractive tool for fast measurements with high sensitivity and high resolution [1]. One approach in dual-comb spectroscopy is to rely on fully passive dual-output (dual-comb) laser systems. The most powerful oscillator among these systems is a dual-comb thin-disk oscillator providing MW-peak power level directly. High output powers immensely simplify nonlinear frequency conversion and this ultimately paves a way towards dual-comb XUV spectroscopy.

Here, we present nonlinear frequency conversion into green spectral region and systematically compare it with the measurements at the fundamental wavelength of 1 µm. Real-time measurements simultaneously at both wavelengths are presented (see figure 1b). This is possible thanks to the excellent mutual stability of our dual-comb oscillator. The high output power of our laser source facilitates frequency doubling of our 1 µm and even further conversion UV. Simultaneous measurement of iodine (515 nm) and acetylene (1030) helps to evaluate the jitter and noise characteristics at both wavelengths.

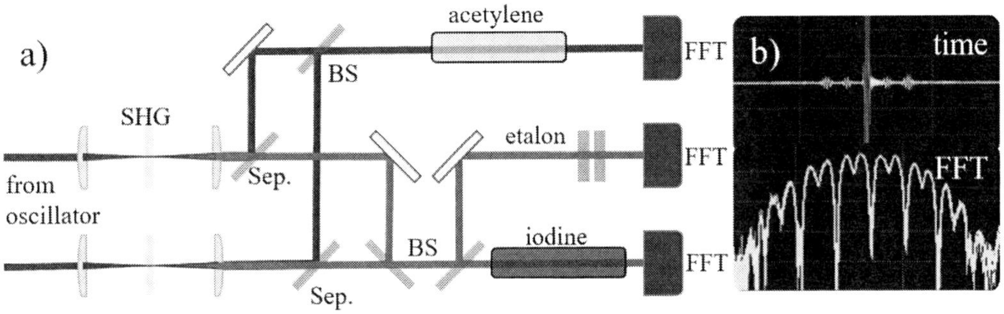

Fig. 1 a) Measurement setup. The 1030 nm IR-beams originated from the dual-comb thin-disk oscillator [2] enter the setup from the left. Both beams are passing identical SHG-stages before the 515 nm radiation is taken apart with separators (Sep.). Both, the IR and green beams are overlapped with beam splitters. The IR superimposed beam passes an acetylene cell and hits the photodiode (PD). The green superimposed beam passes an iodine cell and/or a Fabry-Pérot-etalon before beam measured with the oscilloscope b) A real-time measurement with oscilloscope Keysight DSOS254A. The yellow upper curve shows the time domain signal and the bottom curve the corresponding FFT with down-converted absorption dips induced by the R-branch of acetylene's $3\nu_3$band.

The dual-comb measurement setup is illustrated in figure 1a. The dual-comb laser sources are Yb:YAG based thin-disk 60 MHz-oscillators emitting 1.2 MW and 15 W average power pulses with 180 fs duration and 5.5 nm FWHM at 1030 nm [2]. Thin BBOs with 100 µm thickness are chosen. After separating green from IR and overlapping the beams, the spectrally separated dual-combs are guided through the acetylene and iodine cells. Afterwards the signal gets detected with photodiodes and oscilloscope. The dual-comb spectrometer stays within the necessary mutual stability range avoiding aliasing of the dual-comb spectra within two-hour measurement time. For a recorded 650 ms long oscilloscope trace, containing 143 acetylene and 143 iodine interferograms each, the difference of repetition rates is 217.14 Hz for both measurements. It's jitter was measured to be 26.4 mHz for acetylene and 26.5 mHz for iodine. These two values are nearly the same showing that there is no additional jitter accumulation due to the frequency doubling. The radio frequency jitter of the FFTs can easily be digitally corrected. Iodine has an extremely fine and complex absorption structure which makes it a perfect sample for high resolution spectroscopy studies. DCS of iodine with a frequency-doubled fiber laser has been reported before [3], however, with active, adaptive sampling technique and 3 mW average power. Our next step is the extension and first DCS measurements in UV spectral domain.

References

[1] I. Coddington, N. Newbury, and W. Swann, "Dual-comb spectroscopy," Optica 3, 414-426 (2016).
[2] K. Fritsch, T. Hofer., et al. Dual-comb thin-disk oscillator. Nat Commun 13, 2584 (2022).
[3] T. Ideguchi, A. Poisson, G. Guelachvili, T. W. Hänsch, and N. Picqué, "Adaptive dual-comb spectroscopy in the green region," Opt. Lett. 37, 4847-4849 (2012).

 https://doi.org/10.1051/epjconf/202226701050

Towards Carrier-Envelope Phase Stabilization of a 110 MW Thin-Disk Oscillator

Yasmin Kopp[1], Semyon Goncharov[1], Kilian Fritsch[1] Oleg Pronin[1]

1. Helmut Schmidt University, Holstenhofweg 85, 22043 Hamburg, Germany

Deep ultraviolet (UV) and extreme ultraviolet (XUV) spectral regions are of high interest for precision spectroscopy and metrology applications. For example, the 1S-2S transition of He^+ ions requires a coherent deep UV source around 60.8 nm. Another appealing application is the possibility to create ultra-precise nuclear clocks based on the transition at approximately 150 nm in Thorium-229 [1].

To reach this spectral region one straightforward approach can be frequency up-conversion via high harmonic generation in a gas jet, driven by an existing near-infrared laser. An ideal starting point for this approach can be a home-built carrier-envelope phase (CEP) stabilized thin-disk oscillator [2], delivering 110 MW output peak power at 14 MHz repetition rates. For high harmonic generation, it is necessary to achieve pulses from the oscillator with a low intensity noise. Additionally, to efficiently reach the spectral region, the laser is required to be CEP and repetition-rate stabilized [3]. So far, first steps towards the stabilization with an intra-cavity acousto-optic modulator were performed and the detection of the carrier envelope offset frequency was implemented.

Here, the detection of the carrier envelope offset frequency is demonstrated. First, 12.5 W of the oscillators output power were sent through a multipass-cell (Fig.1 (a)) to spectrally broaden and temporally compress the pulses down to 36 fs. Then, 1 W of the cell output power was used for the generation of an octave spanning spectrum in a 16 cm long photonic crystal fiber (NKT-Photonics LMA-PM-5) with an efficiency of 50 %. The resulted white light was coupled into a f-2f interferometer (Fig.1 (a)). To achieve a temporal overlap, the interferometer arm, covering the blue part of the spectrum, was delayed. The fundamental wavelength at 1440 nm was frequency doubled in a beta barium borate (BBO) crystal. The beat signal was detected with a highly sensitive avalanche photodiode and optimized up to a magnitude of 30 dB (Fig.1 (b)), measured with an RF spectrum analyzer and a resolution bandwidth of 10 kHz. As a next step towards CEP stabilization of the thin-disk oscillator, this beat signal will be send to a phase locked loop including a digital phase detector and PI^2D controller.

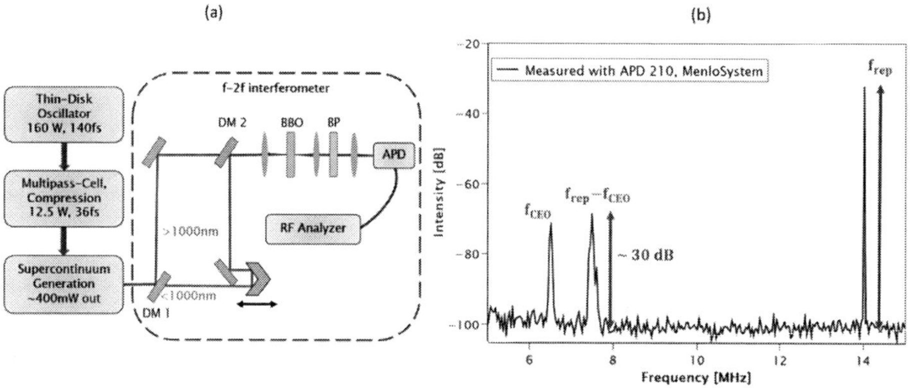

Fig. 1 **(a):** Schematic setup for the f-to-2f interferometer. DM: Dichroic mirror, BBO: Beta-Barium-Borate nonlinear crystal, BP: Bandpass filter at 720 nm, APD: avalanche photo-diode. **(b):** Beat signal and repetition rate with a resolution bandwidth of 10 kHz.

The detection of the beat signal was a first step for CEP stabilisation of the 110 MW thin-disk oscillator. This heterodyning beat signal remained reproducibly and well pronounced. The magnitude of the beat signal needs to be improved to a level of 40 - 50 dB. The stabilization via intra-cavity loss modulation through an acousto-optic modulator and a feedback loop is a work in progress. The CEP stabilization of the thin-disk oscillator paves the way towards the first directly oscillator driven XUV frequency comb.

References

[1] L. von der Wense, B. Seiferle, M. Laatiaoui, J. B. Neumayr, H.-J. Maier, H.-F. Wirth, C. Mokry, J. Runke, K. Eberhardt, C. E. Düllmann, N. G. Trautmann, and P. G. Thirolf, "Direct detection of the (229)Th nuclear clock transition," Nature 533, 47–51 (2016).
[2] S. Goncharov, K. Fritsch, and O. Pronin, "110 MW Thin-Disk Oscillator," ArXiv (2022).
[3] T. Südmeyer, C. Kränkel, C. R. E. Baer, O. H. Heckl, C. J. Saraceno, M. Golling, R. Peters, K. Petermann, G. Huber, and U. Keller, "High-power ultrafast thin disk laser oscillators and their potential for sub-100-femtosecond pulse generation," Appl. Phys. B 97, 281–295 (2009).

Line-search FROG algorithm for retrieval of pulses from noisy datasets

Christoffer Krook[1], Koen Claessen[2], Valdas Pasiskevicius[1]

1. *Department of Applied Physics, Royal Institute of Technology, Roslagstullsbacken 21, 106 91 Stockholm, Sweden*
2. *Department of Computer Science and Engineering, Chalmers University of Technology, Rännvägen 6, 412 58 Gothenburg, Sweden*

Ultrashort-pulse solid-state laser systems are widely used today in a number of applications and often include stages of pulse shaping, spectral broadening, recompression, other processing steps. Experiments and applications in the femtosecond regime require reliable methods to characterize and measure the full electric field of the resulting pulses. Frequency-Resolved Optical Gating (FROG) is a common approach. The algorithms to retrieve pulses from measurement data typically employ gradient descent-based approaches, which tend to retrieve an average of the present measurement noise together with the pulse.

Conventional FROG retrieval algorithms using a gradient descent approach use the experimental data to generate a new guess. Traditional algorithms might reach local minima instead of converging further to the global minimum for complicated signals and noisy datasets. Recent a generic optimization algorithm was published, which presents what is best described as a structured random search of the available input space [1]. Here we have adapted it for pulse retrieval from any FROG geometry. This new algorithm, which we call Line-search FROG, generates two complex values in the complex plane surrounding the current field value along a randomly generated directional vector.

These three points can be considered to form a line in the complex plane, where the current field value intersects the line, and the two generated values make up the endpoints. These three points are evaluated by calculating the trace-area normalized FROG error G', after which the algorithm generates a new point on the line between the best two out of the three points and discards the worst point, effectively shrinking the line. This is repeated until a pre-set number of evaluations have been reached and the algorithm replaces the current guess with the best point of the current three, after which it starts over, generating a new directional vector and two new endpoints. If no better point was found during the iteration, the current guess is replaced with the second-best point from the previous iteration, thus "perturbing" the point a little bit, resulting in a reduced sensitivity to local minima. For a more precise description of the algorithm, we refer the reader to [2].

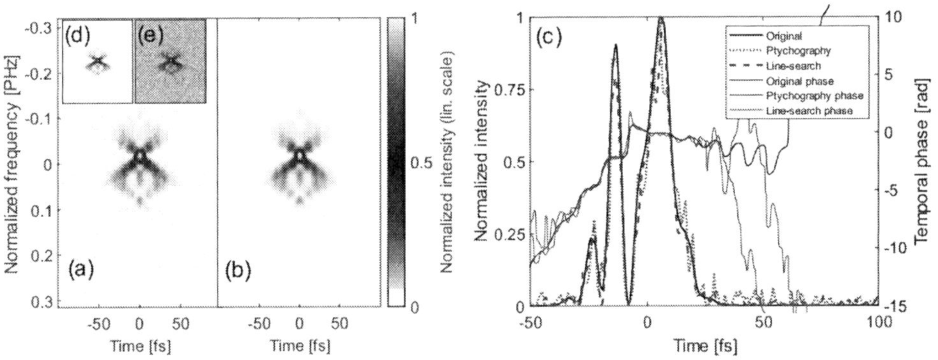

Fig. 1 Trace retrieved by (a) Ptychographic retrieval algorithm and (b) Line-search algorithm together with (c) temporal intensity and phase for retrieved pulses and original the pulse. Inset (d) and (e) show simulated trace before and after adding additive noise amounting to an SNR of 4 dB (the input of the algorithms).

Our results show that the line-search algorithm retrieves pulses with less noise compared to the popular Ptychographic retrieval algorithm [2]. This is evident by comparing both noise in the retrieved traces in Fig. 1 (a) and (b), as well as the noise present in the final pulse in Fig. 1 (c). Unlike for the ptychographic algorithm, each evaluation of the line-search algorithm does not take it closer to the solution, but rather it evaluates the previously mentioned line at several locations, after which it selects the best location. This increases the number of evaluations needed for an acceptable retrieval. For the sake of comparison, both retrieved pulses in Fig. 1 (c) are from 10 000 iterations of each algorithm respectively. The algorithm never sees the evaluation of G', but just the number it returns. Due to this, by changing the evaluation function to match that of for example X-FROG or TG-FROG, the algorithm can work for any FROG geometry. Simulations also show that the algorithm can retrieve pulses from partial datasets the same way as described in [2], as well as blind FROG.

References

[1] Z. Ramezane, K. Claessen, et.al, "Testing Cyber-Physical Systems Using a Line-Search Falsification Method," IEEE Trans. Comput.-Aided Des. Integr. Circuits Syst. doi: 10.1109/TCAD.2021.3110740 (2021)
[2] P. Sidorenko, O. Lahav, et. al, "Ptychographic reconstruction algorithm for frequency-resolved optical gating: super-resolution and supreme robustness," Optica 3, 1320-1330 (2016)

EPJ Web of Conferences 267, 01052 (2022) https://doi.org/10.1051/epjconf/202226701052
EUROPHOTON 2022

High repetition rate, low noise and wavelength stable OPCPA laser system with highly efficient broadly tunable UV conversion for FEL seeding

Tino Lang[1], Mehdi M. Kazemi[1], Jiaan Zheng[1], Samuel Hartwell[1], Nhat-Phi Hoang[1], Eugenio Ferrari[1], Enrico Allaria[1], Lucas Schaper[1] and Ingmar Hartl[1]

1. Deutsches Electron Synchrotron DESY, Hamburg, Germany

High energy tunable ultrafast UV pulses are of great interest for a variety of applications. Within the FLASH2020+ project the FLASH VUV/XUV FEL facility at DESY (Hamburg, Germany) is currently undergoing a major upgrade to become the first high repetition rate, fully coherent FEL light source worldwide [1]. To reach this goal, one of the two in parallel operated FEL branches will be seeded at a fixed wavelength at 343 nm in a first step (SEED 1) and tunable between 297 nm to 317 nm in a second step (SEED 2) following the two-color Echo-Enhanced Harmonic Generation (EEHG) scheme [2]. The seed laser system is designed to deliver UV pulse energies > 100 µJ and > 50 µJ for SEED 1 and SEED 2, respectively, and with 6000 pulses in one second (1 MHz pulse train in 600 µs - 10 Hz bursts). In order to exploit the full capabilities of the narrow-band fully coherent FEL pulses for 24/7 scientific user experiments, the seed laser needs to provide broadly tunable, high power UV laser pulses with pulse durations of 50 fs, excellent beam quality and exceptional high short and long-term stability in respect to the seeding wavelength (< 2e-4), pulse – pulse energy (< 2%) and pointing jitter (< 20 µrad). Altogether, the requirements on the laser system are beyond state-of-the-art.

We will present the concept as well as the first experimental results of our novel high-power seed laser system based on a 5 kW Inno-Slab CPA pump laser system, optical parametric chirped pulse amplification (OPCPA) and a highly efficient UV conversion scheme. An extensive numerical study based on a 3+1 dimensional start-to-end simulation code (chi3D) allows for predictions of system performance in terms of output power, tunability, beam quality and stability in respect to the measured input parameters and respective statistical fluctuations and systematic in-burst sweeps. Fig. 1 shows both, the schematic setup of all sub-sequent nonlinear conversion stages within the laser system and at the same time the numerical simulation blocks. The model includes all nonlinear second order direct and cascaded conversion processes as well as self-phase modulation and self-focusing. The performance of the pump laser system and the super-continuum generation via filamentation in a YAG-crystal are input parameters for the simulation. The theoretical results are confirmed by first experimental studies showing an out-standing high broadband (50 fs transform limit) OPCPA-Signal to UV conversion efficiency of up to 68 % and excellent beam quality. Furthermore, the numerical predicated improvement of the pulse-to-pulse stability compared to the OPCPA output stability could be experimentally reproduced.

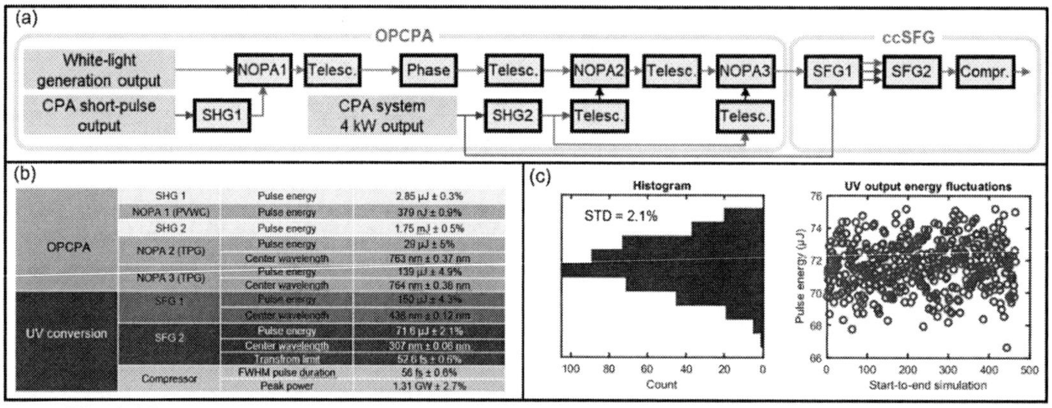

Fig. 1 (a) Simulation building blocks including all second order nonlinear conversion stages as well as SPM, self-focusing and propagation and imaging effects; (b) Simulations results showing the rms value and its corresponding standard deviation of each sub-sequent conversion stage; (c) Exemplary, all start-to-end simulation results and the corresponding histogram in case of UV output pulse energy

The insides of this study had major impact on the conceptual design of the laser system, especially the dispersion concept and the best implementation of remote controls, such as power attenuation and fast wavelength control, etc.

References

[1] M. Beye, Ed., FLASH2020+ : Making FLASH brighter, faster and more flexible : Conceptual Design Report. Deutsches Elektronen-Synchrotron, DESY, Hamburg, 2020. DOI: 10.3204/PUBDB-2020-00465

[2] L. Schaper, ed.al, "Flexible and Coherent Soft X-ray Pulses at High Repetition Rate: Current Research and Perspectivesal" Appl. Sci. 2021, 11, 9729. https://doi.org/10.3390/app11209729

Towards a monolithic, multi-gigahertz mode-locked Ti:Sa laser

Torben Fiehler[1], Ulrich Wittrock[1]

1. Münster University of Applied Sciences, Stegerwaldstr. 39, 48565 Steinfurt, Germany

Multi-gigahertz, low-noise ultrafast lasers have been subject to considerable research interest. They are used in frequency metrology, optical spectroscopy [1] and calibration of astronomical spectrographs [2]. Titanium-doped sapphire (Ti:Sa) is a suitable gain medium due to its broad gain bandwidth. To this day, the highest repetition rate of a Ti:Sa laser is 10 GHz [3]. We present a monolithic design of a fundamentally mode-locked Ti:Sa laser for repetition rates between 30 GHz and 300 GHz.

The laser crystal is a Ti:Sa disk with a thickness in the range of 0.3 mm to 3 mm and 10 mm diameter. Both optical surfaces are planar and coated with dispersion-compensating dielectric coatings. Stable laser operation is achieved by thermal lensing due to absorption of the pump beam. For homogeneous pumping, the medium becomes a gradient index lens whose fundamental eigenmode is a TEM_{00}-mode with a constant diameter along the propagation axis. We want to use soft-aperture Kerr-lens mode-locking. Since the gain bandwidth of Ti:Sa exceeds several tens of nanometers, up to a hundred longitudinal modes can be sustained even at a repetition rate of 300 GHz. Such a monolithic mode-locking scheme allows for low pulse timing jitter and constant pulse repetition rates since there will be no vibrations of resonator mirrors or air turbulence inside the resonator.

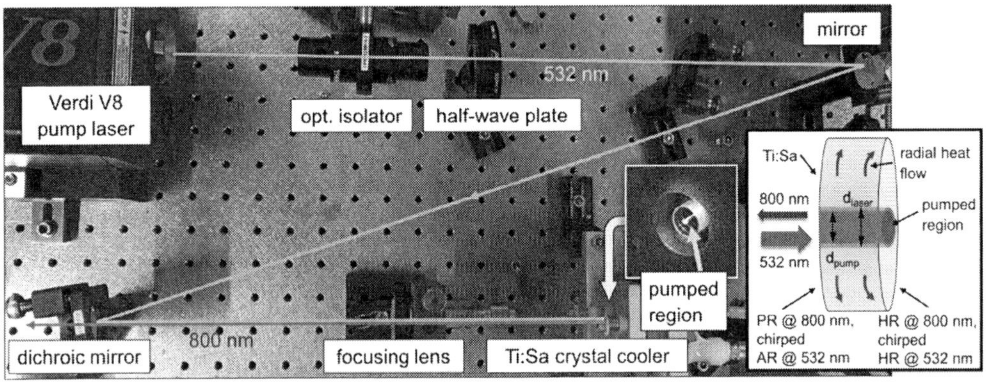

Fig. 1 Experimental pumping setup of the Ti:Sa crystal. The schematic illustration of the monolithic mode-locked Ti:Sa laser shows the coatings on both end faces and the predominantly radial heat flow.

A first batch of 3 mm Ti:Sa crystals was coated and tested. The pump source is a frequency-doubled stabilized single-frequency $Nd:YVO_4$ laser at 532 nm. With an incident pump power of up to 8.5 W and 60% absorption, output powers up to 1.8 W with power stability of $\leq$ 0.04% rms over one hour and 10 nm bandwidth are produced. The interferometric autocorrelation shows a contrast of just 3:1 instead of 8:1 which indicates that we are not seeing mode-locked pulses [4]. The laser beam was then sent into a pulse stretcher introducing 10^4 fs^2 and later 10^6 fs^2 of group delay dispersion and then measured again by the autocorrelator. The measured autocorrelation trace did not change. This suggests that we have not achieved mode-locking but are simply seeing white light interference of the laser light.

Our laser has 10 nm bandwidth in continuous wave operation because for a smaller bandwidth spatial hole burning in the gain regions close to the mirror would produce undepleted gain. Solition mode-locking is only possible if sufficient dispersion compensation by the coatings is achieved over an even larger bandwidth. Mode-locking might also be prevented by frequency pulling due to gain modulation which is caused by the standing wave of the pump beam. An improved coating design is currently being applied to a batch of 1 mm thick Ti:Sa crystals. During the conference, results of experiments with these 1 mm thick crystals will also be presented.

The project is funded by Deutsche Forschungsgemeinschaft (DFG) (Project number 462081165). We thank Clara Saraceno, Günther Steinmeyer and Evgeny Gurevich for helpful discussions.

References

[1] S. M. Link, D. J. H. C. Maas, D. Waldburger, and U. Keller, "Dual-comb spectroscopy of water vapor with a free-running semiconductor disk laser", Science 356, 1164 (2017).
[2] T. Udem, R. Holzwarth, and T. W. Hänsch, "Optical frequency metrology," Nature 416, 233–237 (2002).
[3] A. Bartels, D. Heinecke, and S. A. Diddams, "Passively mode-locked 10 GHz femtosecond Ti:sapphire laser," Opt. Lett. 33, 1905-1907 (2008).
[4] S. Toenger, R. Mäkitalo, J. Ahvenjärvi, P. Ryczkowski, M. Närhi, J. M. Dudley and G. Genty, "Interferometric autocorrelation measurements of supercontinuum based on two-photon absorption," J. Opt. Soc. Am. B 36, 1320-1326 (2019).

8.7-W average power femtosecond Ho:CALGO bulk laser at 2.1 μm

Weichao Yao[1], Yicheng Wang[1], Sergei Tomilov[1], Shahwar Ahmed[1], Christoph Liebald[2], Daniel Rytz[2], Mark Peltz[2], Volker Wesemann[2], and Clara J. Saraceno[1]

[1]Photonics and Ultrafast Laser Science, Ruhr Universität Bochum, Universitätsstraße 150, 44801 Bochum, Germany

[2]Electro-Optics Technology GmbH, Struthstraße 2, 55743 Idar-Oberstein, Germany

High-power ultrafast solid-state lasers directly emitting in the 2 μm spectral region are an important research topic because of growing demands in spectroscopy, polymer processing, and for efficiently driving secondary sources from the THz to the XUV. So far, Cr:ZnS/Se have mostly dominated the landscape of results in this area, achieving power levels up to the watt-level with <100 fs [1]. Ho^{3+}-doped laser gain media are another attractive alternative with largely unexplored potential. In fact, the requirements for high thermal conductivity and a broad gain spectrum is known to be a challenge, and few options for broadband hosts have been explored. In this context, rare-earth doped disordered $CaLnAlO_4$ (Ln stands for Gd/Y/Lu) crystals are known for high thermal conductivity as well as broad gain profile in both 1 μm and 2 μm regime [2, 3]. However, to date, in addition to a few CW laser results based on Ho:$CaYAlO_4$ [4], no reports of mode-locking were made based on Ho:$CaLnAlO_4$. In this paper, we present first mode-locking results with Ho:$CaGdAlO_4$ (Ho:CALGO), reaching the highest average power mode-locked laser so far achieved in the bulk geometry in the 2-3 μm wavelength region.

Our experimental setup is presented in Fig. 1(a). A Brewster-cut 15 mm-length Ho:CALGO crystal (3.1 at.% doping concentration) was used to ensure a double-pass absorption ~80 % and access to π-polarization. The laser was in-band pumped by a single-mode 1940-nm Tm-fiber laser with a pump beam radius of 93 μm (sagittal) × 175 μm (tangential) in the crystal. TEM_{00} mode radius in the crystal was 95 μm (sagittal) × 180 μm (tangential). In CW wave operation (using an HR mirror and an OC as shown in Fig. 1(a)), we could reach up to 10.8 W of output power at an absorbed pump power of 22.8 W, corresponding to an o-o efficiency of 47.4%, as shown in Fig. 1(b). To ensure soliton mode-locking, in addition to the -1800 fs^2 of round-trip group delay dispersion (GDD) from the CALGO crystal, additional -15000 fs^2 GDD was introduced by multiple bounces on dispersive mirrors. A commercial GaSb based SESAM (RefleKron Ltd.) working at 2.1 μm with a saturable fluence of 10 $\mu J/cm^2$ and a modulation depth of 0.22% was used to start and stabilize the mode-locking.

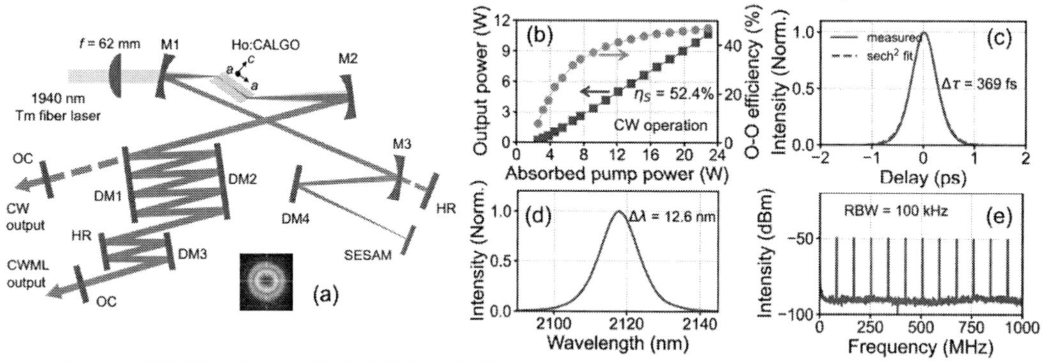

Fig. 1. (a) SESAM mode-locked Ho:CALGO laser. DM1-4 dispersive mirrors. Inset: far-field beam shape during mode-locking at 8.7-W average power. (b) CW operation. (c) Intensity autocorrelation trace with sech² fit. (d) Laser spectrum. (e) RF spectra with 1 GHz scanning range.

Applying an output coupler with 10% transmission, mode-locking was stable up to 22.8-W absorbed pump power with a maximum average output power of 8.7 W, corresponding a high optical-to-optical efficiency of 38.1%. Figure 1(c)-(d) show the laser performance at the maximum output power. Laser wavelength was centered at 2118 nm with 12.6-nm full-width half maximum bandwidth. The obtained pulses have a pulse duration of 369 fs assuming a sech² pulse shape. Time-bandwidth product of the mode-locked pulses was calculated to be 0.311, indicating near transform-limited soliton pulses. The radio frequency (RF) spectra in Fig.1 (d) indicates 84.4 MHz repetition rate resulting in pulse energy of 103 nJ. In this experiment, we were limited to reach higher powers and shorter pulses by the low modulation depth of the SESAM, giving us a straightforward path to improve these results to >10 W and <100 fs level.

References

[1] S. B. Mirov, V. V. Fedorov, D. Martyshkin, I. S. Moskalev, M. Mirov, and S. Vasilyev, "Progress in mid-IR lasers based on Cr and Fe-doped II-IV chalcogenides," IEEE J. Sel. Top. Quantum Electron. **21**(1), 1601719(2015).

[2] Y. Wang, P. Loiko, Y. Zhao, Z. Pan, W. Chen, M. Mero, X. Xu, J. Xu, X. Mateos, A. Major, M. Guina, V. Petrov, and U. Griebner, "Polarized spectroscopy and SESAM mode-locking of Tm,Ho:CALGO," Opt. Express **30**(5), 7883-7893(2022).

[3] S. Ricaud, A. Jaffres, K. Wentsch, A. Suganuma, B. Viana, P. Loiseau, B. Weichelt, M. Abdou-Ahmed, A. Voss, T. Graf, D. Rytz, C. Hönninger, E. Mottay, P. Georges, and F. Druon, "Femtosecond Yb:$CaGdAlO_4$ thin-disk oscillator," Opt. Lett. **37**(19), 3984-3986(2012).

[4] D. Zhou, J. Di, C. Xia, X. Xu, F. Wu, J. Xu, D. Shen, T. Zhao, A. Strzęp, W. Ryba-Romanowski, and R. Lisiecki, "Spectroscopy and laser operation of Ho:$CaYAlO_4$," Opt. Mat. Express **3**(3), 339-345(2013).

EPJ Web of Conferences **267**, 01055 (2022)
EUROPHOTON 2022 https://doi.org/10.1051/epjconf/202226701055

Energy scaling of multi-pass cells for nonlinear optics

Victor Hariton[1,2], Kilian Fritsch[1] and Oleg Pronin[1]

1. Helmut Schmidt University, Holstenhofweg 85, 22043 Hamburg, Germany

2. Instituto Superior Técnico, Universidade de Lisboa, Av. Rovisco Pais 1, 1049-001 Lisboa, Portugal

Nonlinear spectral broadening based on self-phase modulation (SPM) followed by pulse compression is a well-established approach in high average power and ultrashort pulse duration systems [1,2]. One exemplary implementation uses free space propagation through a nonlinear medium (solid or gas) in a multipass arrangement [3]. Scaling this method beyond the 10 mJ-range is mainly limited by three detrimental effects: excessive self-focusing, laser-induced mirror damage and gas ionization [4]. Dominantly multipass cells are realized with a Herriott-cell (HC) arrangement comprising two concave mirrors (cav-cav). Here we suggest a different approach, a concave-convex geometry (cav-vex). No foci are present in these multi-pass cells, mitigating the ionization problem. The decreased intensity permits folding of the arrangement, e.g., as in fig 1a), which reduces the required footprint for practical applications. Based on these properties, we see potential to spectrally broaden 1 J picosecond pulses in a setup with less than 6 m length.

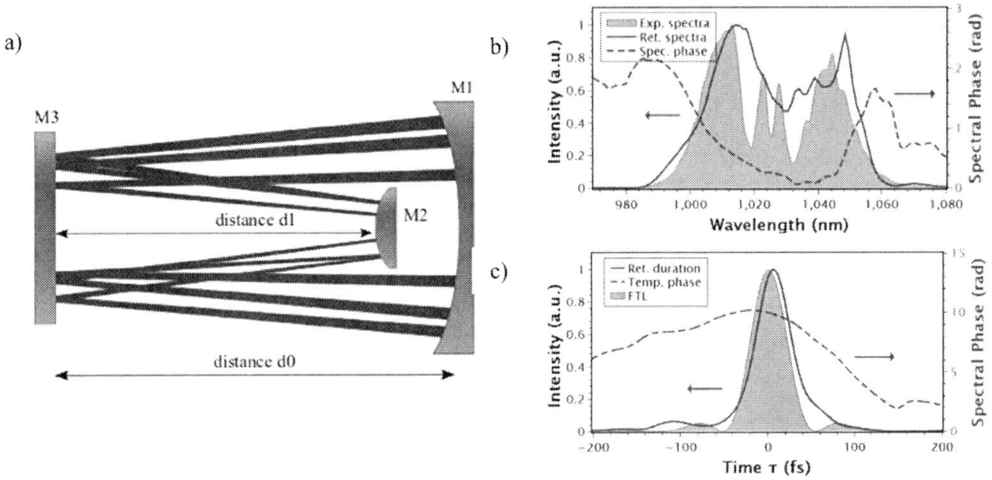

Fig. 1 a) Folding geometry for a convex-concave nonlinear multipass cell. b) Measured (shaded area) and FROG-retrieved (solid line) output spectrum and spectral phase (dashed line) for the Concave – Convex configuration at input energy of 15 μJ. The FROG error is 7×10^{-3} on a 256 × 256 grid. c) FROG-retrieved temporal profile (solid line), temporal phase profile (dashed line) and FTL reference (shaded area). The measured duration of 53 fs with FTL of 49 fs.

To prove the feasibility of this novel concept, we designed a HC broadening setup based on a cav-vex mirror arrangement and a fused silica plate for the nonlinear broadening medium. Pulses with 15 μJ of energy and 260 fs duration, from a commercial laser (Light Conversion PHAROS), are spectrally broadened to a spectrum spanning from 990 nm to 1070 nm (Fig. 1b, shaded area) with a corresponding Fourier-transform limit of 49 fs, which corresponds to a compression factor of 5.5. Compressed pulses are measured using a frequency-resolved optical gating device (APE GmbH), obtaining a pulse duration of 53 fs FWHM (Fig. 1c, shaded area) with a transmission efficiency of 91 %. The concave-convex arrangement delivers excellent spatio-spectral homogeneity with a weighted-average spectral overlap integral exceeding 90 % [3], as good as concave-concave cells. In conclusion, these findings show that using the convex-concave configuration is a viable option for external broadening and compression for a wide range of pulse energy, average power and durations. Moreover, they allow excellent performance in a reduced footprint, avoiding ionization of the media and the damage of the cell mirrors. As a result, the concept should allow energy scaling towards 1 J-level and TW-level input peak powers.

References

[1] J. Schulte, et al., "Nonlinear pulse compression in a multi-pass cell," Optics letters 41, 4511–4514 (2016).
[2] K. Fritsch et al., All-solid-state multipass spectral broadening to sub-20 fs. Optics Letters, 2018. 43(19): p. 4643-4646.
[3] J. Weitenberg, et al., "Nonlinear Pulse Compression to Sub-40 fs at 4.5 μJ Pulse Energy by Multi-Pass-Cell Spectral Broadening," IEEE J. Quantum Electron. 53, 1–4 (2017).
[4] M. Kaumanns, et al.,"Spectral broadening of 112 mJ, 1.3 ps pulses at 5 kHz in a LG_{10} multipass cell with compressibility to 37 fs " Optics letters 46, pp. 929-932 (2021).

EPJ Web of Conferences **267**, 01056 (2022) https://doi.org/10.1051/epjconf/202226701056
EUROPHOTON 2022

Pulse broadening and compression at 515 nm in a multi-pass cell

Victor Hariton[1,2], Kilian Fritsch[1] and Oleg Pronin[1]

1. Helmut Schmidt University, Holstenhofweg 85, 22043 Hamburg, Germany

2. Instituto Superior Técnico, Universidade de Lisboa, Av. Rovisco Pais 1, 1049-001 Lisboa, Portugal

Spectral broadening and compression is a powerful technique to generate short pulses around 1 μm [1,2]. Extending this approach to different wavelengths would be highly beneficial. For instance, increasing demand for coherent and ultrafast laser sources in the 250–530 nm (UV to visible) spectral range is rising for industrial and scientific applications [3]. In this work, we propose shortening the pulses directly in green to not compromise the frequency doubling efficiency. We use a traditional BBO crystal for efficient second-harmonic generation at a moderate (>250 fs) pulse duration, from 1030 nm to 515 nm, followed by a spectral broadening and compression scheme in a Herriot-type cell (HC), using a bulk material as the nonlinear medium.

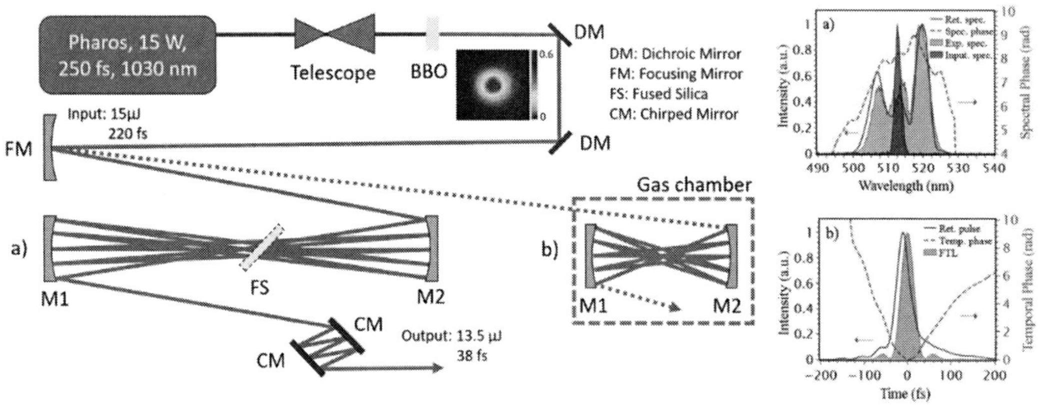

Fig. 1 Left: Experimental setup for pulse broadening and compression in the 515 nm spectral region. The 1030 nm IR-beams originated from a commercial laser (Light Conversion PHAROS) are frequency-doubled in a BBO crystal and subsequently the 220 fs pulses are broadened in a) a fused silica plate or b) in a pressurized environment containing different gases. Right: a) measured (shaded area) and FROG-retrieved (solid line) output spectrum and spectral phase (dashed line) for the solid-state broadening scheme at input energy of 15 μJ. b) FROG-retrieved temporal profile (solid line), temporal phase profile (dashed line), and FTL reference (shaded area). The measured duration was 38 fs with an FTL of 35 fs.

The driver laser (Light Conversion PHAROS) delivers 15 W average power at 1 μm, emitting 250 fs pulses with a pulse energy of up to 200 μJ at a repetition rate of up to 1.0 MHz. After the conversion stage, the pulses are spectrally broadened and compressed in a multi-pass cell yielding sub-40 fs pulses with over 20 nm bandwidth around 515 nm. The efficiency of spectral broadening and compression exceeds 90 %. These results represent the first implementation of a multi-pass spectral broadening and compression in the green spectral regions and pave the way for similar experiments in UV.

References

[1] J. Weitenberg et al., Multi-pass-cell-based nonlinear pulse compression to 115 fs at 7.5 μJ pulse energy and 300 W average power. Optics Express, 2017. 25(17): p. 20502-20510.
[2] K. Fritsch et al., All-solid-state multipass spectral broadening to sub-20 fs. Optics Letters, 2018. 43(19): p. 4643-4646.
[3] M. Gu, D. Bird, D. Day, L. Fu, and D. Morrish, Femtosecond Biophotonics: Core Technology and Applications (Cambridge University Press, New York, 2010).

Smart and agile 88 W Yb-fiber frequency comb laser

Sarper Salman[1,2,3], Mingqi Fan[1], Henrik Tünnermann[1], Prannay Balla[1,2,3], John Darvill[1], Dominic Laumer[1], Vito F. Pecile[4], Jakob Fellinger[4], Valentina Shumakova[4], Christoph Mahnke[1], Yuxuan Ma[1], Christian Mohr[1], Oliver H. Heckl[4], Christoph M. Heyl[1,2,3] and Ingmar Hartl[1]

1 Deutsches Elektronen-Synchrotron DESY, Notkestraße 85, 22607 Hamburg, Germany

2 Helmholtz-Institute Jena, Fröbelstieg 3, 07743 Jena, Germany

3 GSI Helmholtzzentrum für Schwerionenforschnung GmbH, Planckstrasse 1, 64291 Darmstadt, Germany.

4 University of Vienna, Faculty of Physics, Faculty Center for Nano Structure Research, Christian Doppler Laboratory for Mid-IR Spectroscopy and Semiconductor Optics, Boltzmanngasse 5, A-1090 Vienna, Austria

Optical frequency comb lasers are used in many applications in science and technology. For most applications output power levels around a few watts are sufficient. However, some applications require significant higher power ranges, for example to drive extreme nonlinear processes in gases for extreme ultraviolet (XUV) frequency comb generation. Although, efficient power scaling of fiber-frequency combs to the 100 W level was demonstrated more than 10 years ago [1] and recently kW average power levels on CEP (carrier-envelope phase) stabilized fiber laser systems were achieved [2], there are still only a few high-power frequency comb systems reported to date [3], typically requiring a team of experienced laser scientists for operation.

Here we demonstrate the first steps towards closing the gap in versatility and user-friendliness of high-power optical frequency comb systems by introducing digital controls for pulse shaping, dispersion management and comb-stabilization on a 65 MHz Yb-fiber frequency comb system delivering more than 88 W output power.

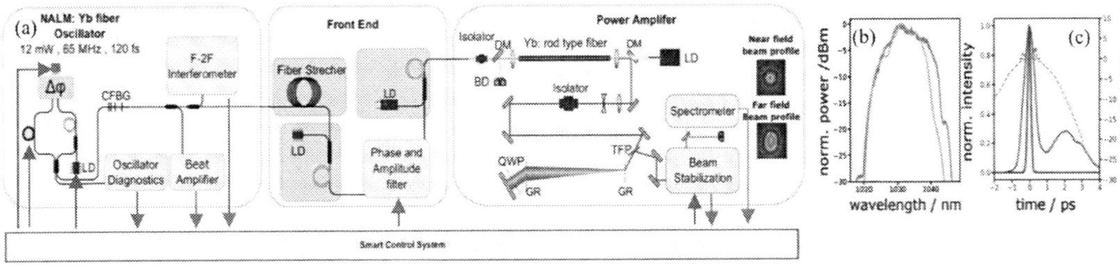

Fig. 1 (a) Schematic of the experimental set-up. LD: Laser diode, TFP: Thin Film Polarizer, GR: Grating, CFBG: chirped fiber Bragg grating, BD: beam dump, QWP: quarter wave plate; (b) measured spectral intensity (green) by optical spectrum analyzer (OSA) and retrieved spectral intensity (orange) by frequency resolved optical gating (FROG); (c) retrieved temporal intensity and phase (FROG) before (blue) and after (red) optimizing the phase and amplitude filter.

The laser system shown in Fig. 1(a) is seeded by 65 MHz compact and low noise Yb:fiber nonlinear amplifying loop mirror (NALM) oscillator [4]. For dispersion management, we use a combination of passive fiber and a frequency-domain phase and amplitude filter (FINISAR 1000A) which is sandwiched between two Yb-doped fiber amplifier stages. We employ an Yb-doped rod type fiber (NKT Photonics) for the last gain stage to reach more than 110 W output power before the compression. Brain of the laser system are its digital controls: The entire laser-system is operated by the control system DOOCS developed for DESY's X-ray light sources [5]. For comb stabilization the phase-locked loops are digitally implemented via field-programmable gate arrays (FPGAs) and driven by the open-source python package pyrpl [6]. We also digitally control the frequency-domain phase and amplitude filter. It is not only used for versatile dispersion control to reach the shortest output pulse Fig. 1(c), but also for optimization of the output pulse phase and amplitude for controlled center wavelength shifting in nonlinear multipass cavities, translating our Yb-comb into a wavelength-tunable source covering a spectral range of 1000 nm - 1060 nm [7].

References

[1] A. Ruehl, A. Marcinkevicius, M. E. Fermann, and I. Hartl, Opt. Lett 35, 3015–3017 (2010).
[2] E. Shestaev, S. Hädrich, N. Walther, T. Eidam, A. Klenke, I. Seres, Z. Bengery, P. Jójárt, Z. Várallyay, Á. Börzsönyi, and J. Limpert, Opt. Lett., 45, 6350–6353 (2020).
[3] T. Fortier, E. Baumann, Commun. Phys. 2, 1-16 (2019).
[4] Y. Ma, S. Salman, C. Li, C. Mahnke, Y. Hua, S. Droste, J. Fellinger, A. S. Mayer, O. H. Heckl, C. M. Heyl, and I. Hartl, J. Lightwave Technol. 39, 4431–4438 (2021).
[5] O. Hensler, K. Rehlich, Proceedings, 15th Conference on Charged Particle Accelerators 22-24, 308–315 (1996).
[6] L. Neuhaus, R. Metzdorff, S. Chua, T. Jacqmin, T. Briant, A. Heidmann, P. Cohadon, S. Deléglise., European Conference on Lasers and Electro-Optics and European Quantum Electronics Conference (2017).
[7] P. Balla et al., "Ultrafast serrodyne optical frequency translator", submitted for publication (2022).

High Average Power Nonlinear Pulse Compression in a Gas-filled Multi-pass Cell at 2 μm Wavelength

P. Gierschke[1,3], C. Grebing[1,2], M. Abdelaal[3], M. Lenski[3], J. Buldt[3], Z. Wang[3], T. Heuermann[3,4,5] M. Müller[3,6], M. Gebhardt[3,4,5], J. Rothhardt[1,3,4,5] and J. Limpert[1,3,4,5]

1. Fraunhofer Institute for Applied Optics and Precision Engineering, Albert-Einstein-Str. 7, 07745 Jena, Germany
2. Present address: Active Fiber Systems GmbH, Ernst-Ruska Ring 17, D-07745 Jena, Germany
3. Institute of Applied Physics,Abbe Center of Photonics,Friedrich Schiller University Jena,Albert-Einstein-Str.15, 07745Jena, Germany
4.Helmholtz-Institute Jena, Fröbelstieg 3, 07743 Jena, Germany
5.GSI Helmholtzzentrum für Schwerionenforschung, Planckstraße 1, 64291 Darmstadt, Germany
6.Present address: Institute de Physique Université de Neuchâtel, Avenue de Bellevaux 51, 2000 Neuchâtel, Switzerland

Ultrafast lasers sources in the short-wavelength infrared region (SWIR, 1.5 μm to 3.0 μm wavelength) are of great interest for manifold direct applications, as well as for nonlinear frequency conversion techniques addressing terahertz, mid-infrared or even soft to hard X-ray spectral regions [1-3]. Especially, the so called "water window" spectral region which expands from around 280 eV–530 eV is of utmost interest in high-resolution imaging or SXR-spectroscopy of biological specimen [4]. These and other applications emphasize the demand for scalable, high-flux, table-top coherent soft X-ray sources which can be obtained via high-order harmonic generation (HHG) from SWIR driving lasers. One concept that can provide ultrafast high-average power SWIR pulses is optical parametric chirped-pulse amplification (OPCPA) [5]. Another concept aimed at high-power ultrashort pulse generation in the SWIR is based on thulium-doped fiber chirped-pulse amplification systems (Tm:FCPA). These lasers deliver multi-GW peak powers, multi-100 W average output power with pulses as short as sub-100 fs [6]. Nevertheless, a subsequent pulse shortening according to the above discussion is typically desired.

Traditionally, this is done in waveguides such as noble gas-filled capillaries [7]. Besides fiber-based post-compression schemes, multi-pass cells (MPC) have emerged as an attractive approach for nonlinear post-compression, showing excellent preservation of the input beam quality and high throughput efficiency [8-9]. Until now, MPCs have been rarely demonstrated in the SWIR wavelength regime, especially at high average power levels of several tens of Watt.

We report on the first nonlinear pulse post compression of the output from a Tm:FCPA using a gas-filled Herriott-type MPC. The driving laser system delivers 138 fs pulses (FWHM) and 65 W of average output power at a repetition rate of 300 kHz, centered at a wavelength of 1940 nm. The MPC is filled with krypton gas at 3 bar of pressure and the throughput is as high as 80%. After the beam propagation through the cell, fused silica is employed to post compress the spectrally broadened pulses to a FWHM duration of 35 fs, with an estimated peak power of several GW. For analyzing the post compressed output pulses in detail, we employed an in-house built SHG-FROG device. The corresponding results are depicted in Fig.1. At 300 kHz repetition rate, a record compressed average power of 51 W is demonstrated, which is the highest to date from any SWIR MPC-based nonlinear pulse post compression experiments.

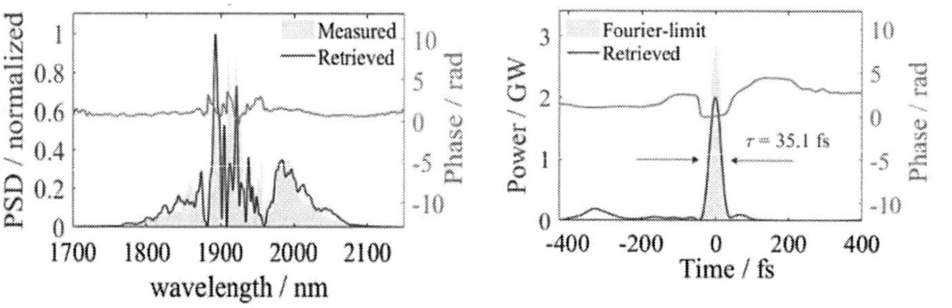

Fig. 1 FROG measurement of the post compressed MPC output.
Left plot, measured and retrieved spectrum. Right plot, retrieved pulse and corresponding Fourier-limit.

This result shows the scalability and enormous potential of the MPC technology in the 2 μm wavelength region, where nonlinear pulse compression at high average power is challenging due to the unfavorable existence of water vapor absorptions [10]. The experiment presented in this contribution is an important development step towards highly application relevant laser-driven nonlinear frequency conversion experiments and further performance enhancement towards mJ-energy, few-cycle laser sources at 2 μm wavelength with >100 W average power.

References

[1] M. Clerici et. al., Phys. Rev. Lett. 110, 253901 (2013).
[2] J. Zhang et. al., Light Sci. Appl. 7, 17180 (2018).
[3] A. Koc et. al., Opt. Letters, 46(2), 210-213 (2021).
[4] Rose, M et. Al., Opt. Express., 26(2), 1237-1254 (2020).
[5] J. Pupeikis et. Al., Optica, 7(2), 168-171 (2020).
[6] T. Heuermann et. al., "Ultrafast Tm-doped fiber laser system delivering 1.65mJ, sub-100 fs pulses at 100 kHz repetition rate" to be published in Opt. letters (2022).
[7] V. Cardin et. Al., Appl. Phys. Lett. 107, 181101 (2015).
[8] M. Hanna et. al., J. Opt. Soc. Am. B 34, 1340 (2017).
[9] A Viotti et. al., Optica 9, 197 (2022).
[10] M. Gebhardt et. Al., Opt. Express 23(11), 13776 (2015).

 https://doi.org/10.1051/epjconf/202226701059

Single-frequency praseodymium doped YLF laser design and operation with extended wavelength coverage in the visible

Paul White[1], Alan J. Kemp[2], Loyd J. McKnight[1]

1. *Fraunhofer Centre for Applied Photonics, Technology and Innovation Centre, 99 George Street, Glasgow, Scotland, UK*
2. *Institute of Photonics, University of Strathclyde, Technology and Innovation Centre, 99 George Street, Glasgow, Scotland, UK*

Single-frequency operation of a diode-pumped praseodymium-doped YLF laser has been demonstrated using an elegant cavity design. Over 100 mW of single-frequency operation has been achieved from 687 nm to 705 nm with one cavity arrangement. This laser system is well suited for applications in atomic cooling and spectroscopy and this laser system targets use in neutral strontium optical clocks[1].

A 5W GaN diode laser was used as a pump laser and was temperature controlled to ensure maximum absorption in the crystal. A combination of cylindrical and spherical beam-shaping optics were used to compensate for varying beam quality and astigmatism between fast and slow axis to result in a near-circular pump spot size of 70um. A V-shaped cavity was carefully designed to ensure maximum output power using a pump-through mirror, 75mm ROC HR mirror and 1% output coupler with coatings selected for operation in the red-NIR wavelength range. Tuning was achieved with a birefringent filter (BRF). Output powers from 200-400 mW were obtained with single-transverse mode multi-longitudinal mode operation.

Single-frequency operation was achieved with careful selection of an uncoated YAG etalon and confirmed using a scanning fabry-perot interferometer. In combination with filter effects from the crystal etalon and BRF single-frequency operation could be achieved without substantially compromising power. Our analysis of the cavity design partially shown in Fig 1 has shown we can compensate for spatial hole burning [2] with appropriate etalon selection for maximum output power. The low-gain from 687-696 nm would otherwise prevent laser operation with this pump power in ring-laser cavities and other cavity designs due to the loss induced from intracavity filter and tuning elements. Our design is carefully optimised to minimise loss whilst providing sufficient filtering. Tuning was achieved from 687 nm to 705 nm as shown in Fig 1. and through tuning of the BRF, etalon, crystal temperature and cavity length any frequency could be selected within that range. Frequency stabilisation to external reference cavities has been achieved with feedback a cavity mirror mounted on a piezo.

The single-frequency power levels across this broad wavelength range provides a versatile platform for many applications in this spectral range in particular for neutral strontium optical clocks. This platform offers advantages over semiconductor sources due to the low-levels of noise from carrier dynamics and the long-lifetime of the gain crystal.

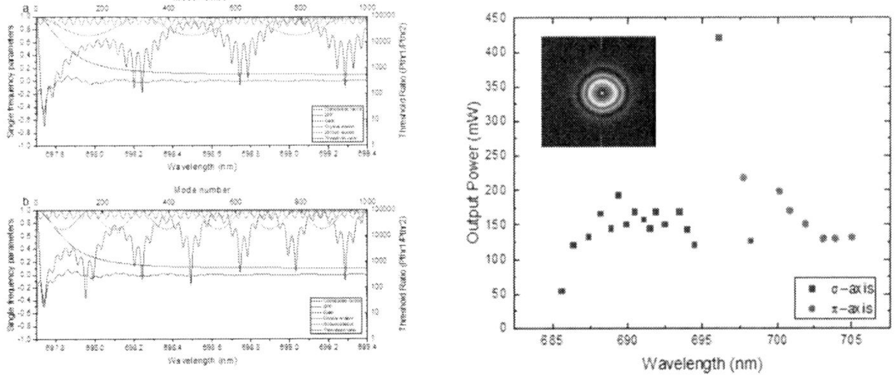

Fig. 1 (left) Analysis of the frequency selective components used to achieve singe-frequency operation and the threshold ratio of the second longitudinal mode compared to the first. (right) measured single-frequency power with beam quality measurement inset (M2 <1.1)

References

[1] A. Sottile, E. Damiano, A. Di Lieto, and M. Tonelli "Diode-pumped solid-state laser platform for compact and long-lasting strontium-based optical clocks", Optics Letters Vol. 44, Issue 3, pp. 594-597 (2019)
[2] J. J. Zayhowski "Limits imposed by spatial hole burning on the single-mode operation of standing-wave laser cavities", Optics Letters Vol. 15, Issue 8, pp. 431-433 (1990)

 https://doi.org/10.1051/epjconf/202226701060

Recent advances in SWIR and MWIR solid-state and fiber sources

Marc Eichhorn[1, 2]

1. Fraunhofer IOSB (Institute of Optronics, System Technologies and Image Exploitation), Gutleuthausstraße 1, 76275 Ettlingen, Germany

2. Institute of Control Systems, Karlsruhe Institute of Technology, Fritz-Haber-Weg 1, 76131 Karlsruhe, Germany

Laser sources in the short-wave infrared (SWIR) around 2 µm wavelength and non-linear converters covering the mid-wave infrared (MWIR) spectral range are important for a large variety of applications like in environmental sensing, detection and ranging, optical free-space communication, optical countermeasures, medical treatments and surgery, and fundamental physics research. Especially recent developments in two-micron pulsed thulium and thulium-holmium doped solid-state and fiber lasers allow for significant average-power and pulse-energy scaling of mid-IR OPOs based on conversion crystals like ZnGeP2, CdSiP2 and OP-GaAs, thereby broadening the scope of various applications. Continuous-wave (cw) sources in the 2 µm range are important for materials processing and communication. Therein, a focus is put on all-fiber designs and robust, if possible self-aligning, laser resonators, which allow for stable and ruggedized designs for industrial applications under harsh environments. The presentation gives an overview of fundamental concepts, recent achievements as well as novel results in the field.

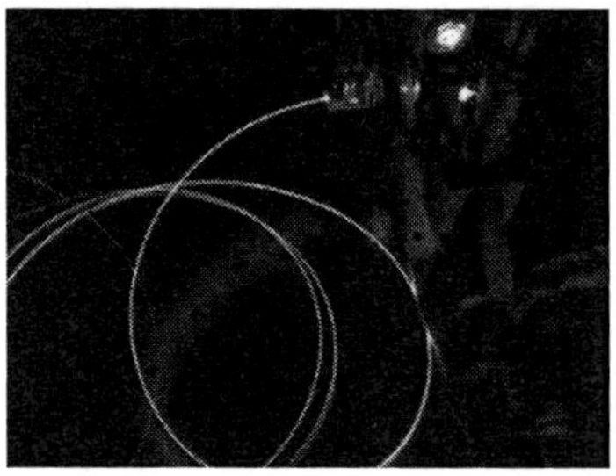

Fig. 1 Thulium-doped fiber laser operating in the 2 µm wavelength range.

Fiber lasers: In the field of SWIR fiber lasers significant achievements have been obtained in all-fiber cw thulium-doped and thulium-holmium co-doped fiber lasers, see fig. 1. By employing a single amplifier stage with a thulium-doped fiber laser, up to 0.94 kW of output power at 58 % slope efficiency and a linewidth less than 0.2 nm have been realized at around 2036 nm. In order to shift the output wavelength further, thulium-holmium co-doped fibers were used to realize a single-oscillator fiber laser with 145 W of cw output power at 2.2 µm, thereby significantly shifting the power limits in this wavelength range. In pulsed operation, record values in pulse energy of 0.96 mJ in Gaussian pulses with a pulse width of 44 ns and a peak power of 20.5 kW were obtained with thulium-doped step-index fibers. The output spectrum shows a narrow FWHM linewidth of < 1 nm.

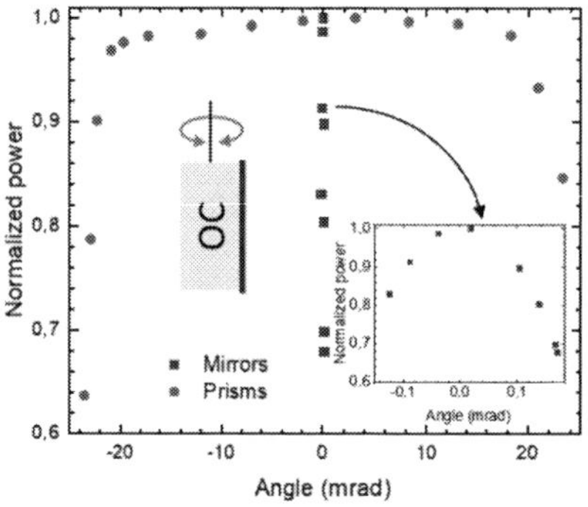

Fig. 2 Alignment sensitivity of a mirror and a Porro-prism resonator of a Ho^{3+}:YAG laser.

Solid-state lasers: A first continuous-wave crossed-Porro-prism laser was recently demonstrated based on a quasi-three-level Ho^{3+}:YAG crystal with output powers > 30 W – an important step to guide future quasi-three-level laser designs which were so far not considered efficiently compatible with Porro-resonators as a result of the reabsorption loss in these materials. As can be seen in fig. 2, an alignment sensitivity up to a factor of 200 lower than the corresponding mirror resonator can be achieved. By using ZGP non-linear crystals, record conversion slope efficiencies of 60 % at 8.2 W average output power could be realized using 2 µm temporally rectangular pulses from a fiber MOPA. This marks a new power scaling of this topology and will allow for further power increase in the near future.

EPJ Web of Conferences **267**, 01061 (2022) https://doi.org/10.1051/epjconf/202226701061
EUROPHOTON 2022

Optical frequency combs from modelocked lasers

Ursula Keller
Physics Department, ETH Zurich, Switzerland

The revolution in optical frequency comb generation started in 1999. This lecture will give an introduction of modelocked frequency comb generation, noise characterization and single-cavity dual-comb generation. I will refer to my new textbook "Ultrafast Lasers" (Springer Verlag 2022) which provides a more comprehensive introduction to fundamental principles and applications.

The frequency comb revolution [1-3] brought two research communities together which had rarely interacted previously. The frequency metrology community was interested in narrowband single-frequency lasers, and the ultrafast laser science community in broadband ultrafast lasers. Both fields were relatively mature around the year 2000, with high-performance single-frequency and modelocked lasers, and with a well-established understanding in both the time and frequency domains in both communities.

This lecture gives a basic introduction to optical frequency comb generation from modelocked lasers, and their noise characterization. For frequency metrology applications the exact position of the carrier (i.e., the electric field below the pulse envelope) needs to be stabilized and controlled. This carrier-envelope offset (CEO) was introduced in 1999 and is referred to as the carrier-envelope offset phase or the carrier-envelope phase. Normally, no locking mechanisms exist between envelope and carrier inside a modelocked laser oscillator, and the relative CEO phase experiences random fluctuations. I will give a basic introduction to optical phase and frequency noise, starting from the basic mathematical principles, CEO phase noise measurement and stabilization techniques based on power spectral densities (PSDs), the connection with intensity noise and timing jitter, integrated noise, and novel measurement techniques for dual-comb noise characterization. The connection between the optical laser spectrum and phase noise is discussed in detail, using the useful concept of the β-separation line. The basic principle of active stabilization and the different laser technologies for optical frequency combs are reviewed.

References

[1] H. R. Telle et al., Appl. Phys. B **69**, 327-332 (1999)
[2] D. J. Jones et al., Science **288**, 635-639 (2000)
[3] A. Apolonski et al., Phys. Rev. Lett. **85**, 740-743 (2000)

Fundamentals of attosecond science

A L'Huillier

Department of Physics, Lund University, Lund, SE-22 100, Sweden

Extreme Ultraviolet light sources based on high-order harmonic generation (HHG) in gases are now used in many areas of science. The radiation consists of a train of extremely short light bursts, in the attosecond range, allowing for outstanding temporal resolution. HHG has opened the field of attosecond science, capturing ultrafast electron dynamics in matter [1]. This presentation will give a short historical perspective on this field of research.

The HHG efficiency depends both on the response of a single atom to a strong laser field and, as in any coherent nonlinear optical process, on the phase matching between the waves emitted by individual atoms in a macroscopic medium. An introduction to the physics of HHG will be given, covering both aspects of the process. The importance of phase matching is illustrated in Figure 1, showing the conversion efficiency of the 23rd harmonic in argon [2]. A good conversion efficiency can be obtained both with a long medium, low pressure and a short medium, high pressure, allowing for versatile setups. The implications of geometrical scaling [3] and wavelength scaling [4] will be discussed.

The short pulse duration and broad bandwidth of attosecond pulses allow the study of photoionization dynamics using laser-assisted photoionization. We will describe in particular the interferometric technique RABBIT (reconstruction of attosecond beating by two-photon transition) method [5], which has been used first for the measurement and characterization of attosecond pulses, then for the determination of photoionization time delays in atoms and molecules [6].

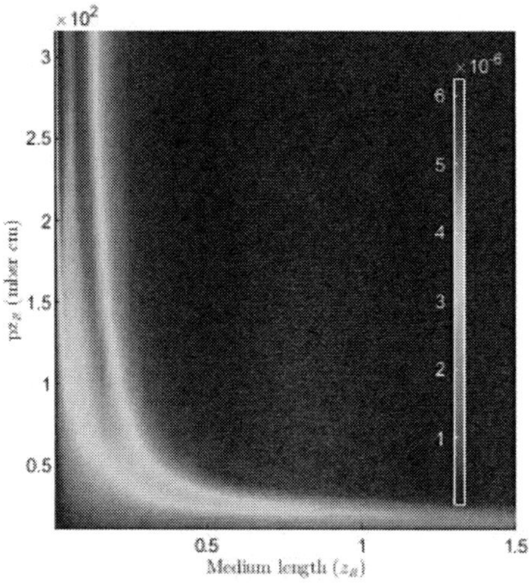

Figure 1. Conversion efficiency of the 23rd harmonic of 800 nm radiation generated in argon at an intensity of 2.5 x10^{14} W/cm^2. z_R denotes the Rayleigh length. Adapted from [1].

References

[1] F. Krausz and M. Ivanov, Rev. Mod. Phys. 81, 163 (2009)
[2] R. Weissenbilder R et al arXiv:2202.08202 (2022)
[3] C. M. Heyl et al., J. Phys. B, 50, 013001 (2017)
[4] B. Sheehy et al., Phys. Rev. Lett. 83, 5270 (1999); T. Popmintchev et al., Science 336, 1287 (2012)
[5] P.-M. Paul et al., Science 292, 1689 (2001)
[6] K. Klünder et al., Phys. Rev. Lett. 106, 143002 (2011)

 https://doi.org/10.1051/epjconf/202226701063

Femtosecond Optical Parametric Oscillators and Frequency Combs

Derryck T. Reid[1]

1. School of Engineering and Physical Sciences, Heriot–Watt University, Edinburgh, EH14 4AS, UK.

Since Burneika's first demonstration in 1972 of a synchronously pumped optical parametric oscillator (OPO) [1], this technology has evolved and matured considerably, opening applications in multiple areas of sensing and metrology. This two-hour session will review a range of topics needed to understand the design and operation of femtosecond optical parametric oscillators and frequency combs based on these devices.

Concepts underpinning femtosecond OPOs will first be introduced. These will comprise discussions of: frequency-domain effects, including phasematching and bandwidth considerations; time-domain effects, including pulse walkaway and dispersion; and feedback, in particular the role of the cavity in the operation of singly- and doubly-resonant OPOs.

Architectures operating from the near- to mid-infrared will then be reviewed according to: the choice of nonlinear gain crystal (e.g. BIBO, BBO, LBO, KTP, PPLN, CSP, OPGaAs, OPGaP); resonator design (linear, ring, fibre feedback); pulse repetition rate (several MHz – multi-GHz); and wavelength coverage (visible – LWIR). As an example, Fig. 1 shows a 10-GHz femtosecond OPO developed at Heriot-Watt University [2].

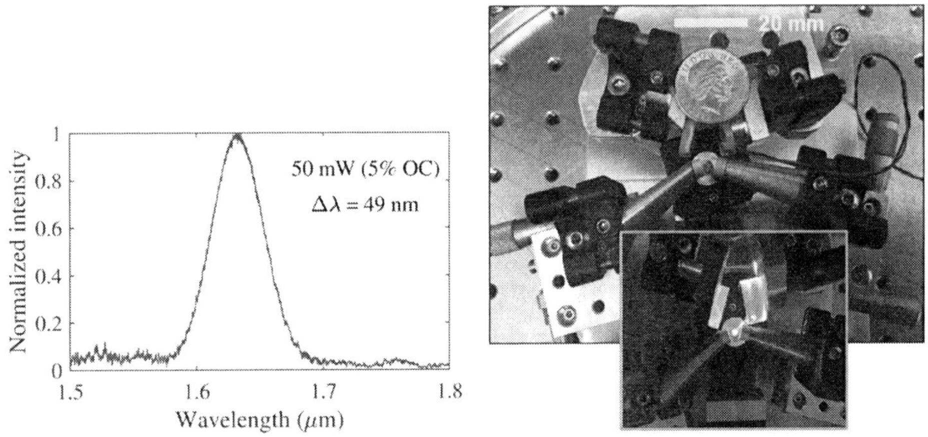

Fig. 1 A doubly-resonant femtosecond OPO pumped by a 10-GHz Ti:sapphire laser and based on the nonlinear material PPKTP. The green light is sum-frequency mixing between the pump and signal pulses.

Techniques for modelling femtosecond OPOs using numerical simulations will be introduced, and sample code supplied. These methods will include classical pulse propagation modelling based on solving the three coupled-wave equations, and also more modern techniques which utilise variations of the generalised nonlinear Schrödinger equation (full-field methods), and which make fewer assumptions about the number and type of nonlinear interactions permitted [3].

Discussion of OPO frequency combs will begin by introducing some fundamental principles of OPO phase control, including the detection and control of the carrier-envelope offset frequency in an OPO. Key applications of OPO frequency combs will then be reviewed, including in frequency metrology, mid-infrared dual-comb spectroscopy, quantum computation and astrophotonics.

The talk will conclude with a short discussion of future perspectives for femtosecond OPOs.

References

[1] K. Burneika et al., "Parametric light amplification and oscillation in KDP with mode-locked pump," IEEE Journal of Quantum Electronics **8**, 574 – 574 (1972).
[2] R. A. McCracken, Y. S. Cheng, and D. T. Reid, "10-GHz Femtosecond Degenerate Optical Parametric Oscillator," in Conference on Lasers and Electro-Optics, OSA Terchnical Digest (online) (Optica Publishing Group, 2018), paper FTh1M.1.
[3] D. T. Reid, "Ultra-broadband pulse evolution in optical parametric oscillators," Opt. Express **19**, 17979-17984 (2011).

EPJ Web of Conferences **267**, 01064 (2022)
EUROPHOTON 2022
https://doi.org/10.1051/epjconf/202226701064

Power Scaling of Solid State Lasers

A. Tünnermann

Fraunhofer Institute for Applied Optics and Precision Engineering
Friedrich-Schiller University Jena, Institute of Applied Physics

Solid-state lasers are attractive sources of coherent radiation for various scientific and industrial applications. in the past 60 years different laser geometries have been developed to overcome challenges in power scaling conserving the beam quality. Rare-earth-doped fiber lasers have emerged as most attractive and power scalable solid-state laser concept due to the outstanding thermo-optical properties of an actively doped fiber. The large ratio of surface to active volume of such a fiber ensures excellent heat dissipation, furthermore the beam quality is defined by the refractive index profile of the active core and is therefore independent on the pump power. Fiber lasers and amplifiers offer a very high single-pass gain and therefore low laser thresholds and efficient diode-pumped operation. Using advanced fiber designs, in continuous-wave and pulse operation output powers exceeding the 10 kW-level with diffraction-limited beam quality have been demonstrated.

However, power and energy scaling of cw and pulsed single-mode fiber lasers and amplifiers is restricted due to nonlinear pulse distortions, which are enforced by the large product of intensity and interaction length inside the fiber core. In addition, transverse mode instabilities are observed which degrade the beam quality emitted by high-power fiber laser systems once that a certain average power threshold has been reached. This sudden degradation of the output beam quality is accompanied by temporal fluctuations of the beam profile. Most recently, strategies have been developed to mitigate or even, ideally, to overcome these limitations. Coherent combination of different fiber amplifier channels enables a further scaling of these parameters.

This contribution presents the physical and technical basics of solid-state lasers and their power scalability. The state of the art of science and technology and prospects for future developments in fiber lasers and amplifiers are reviewed.

Intelligent control of Lasers for Accelerators

Henrik Tünnermann
Deutsches Elektronen-Synchrotron DESY, Notkestraße 85, 22607 Hamburg, Germany

For optimal operation, accelerators and FELs require precise control of their control parameters. Lasers are of critical importance for photocathode, FEL seeding, and probe lasers. We will show our current results and plans to optimize the performance (pulse parameters, fast set-point tuning, stability) of our photocathode and pump-probe lasers using AI methods.

In an accelerator, the laser driving the electron generation in the photocathode is of critical importance. The X-ray FEL research infrastructures in Hamburg (FLASH [1], European XFEL) operate in "burst mode" where a train of high repetition rate X-ray pules is generated every 100ms. This operation mode is very challenging for laser systems since they are not in thermal equilibrium. For example, the pulse energies, pulse shape, the beam pointing, and beam size of the output beam can vary along with the burst. We have developed a feed-forward method to enable control of each of the pulses in the burst. To further improve the FEL pulses the charge profile – especially the emittance – has to be optimized. This can be achieved by shaping of the individual laser pulses. Here the challenge is the large parameter space when shaping the laser pulse as well as the time-consuming simulation of the photocathode. The flow of this shaping procedure is shown in Fig.1. To address these challenges, we are going to use a neural network to speed up calculations.

For the pump-probe lasers used to acquire a snapshot of a chemical reaction with femtosecond resolution, high laser stability is required. During the 100ms gap between X-ray and laser bursts, no laser parameters can be measured and therefore active feedback is of limited use. However, using a deep learning model presents the opportunity to anticipate some of the changes and react in advance. This concept was evaluated on static data and about 30% of timing fluctuations can be predicted in this way.

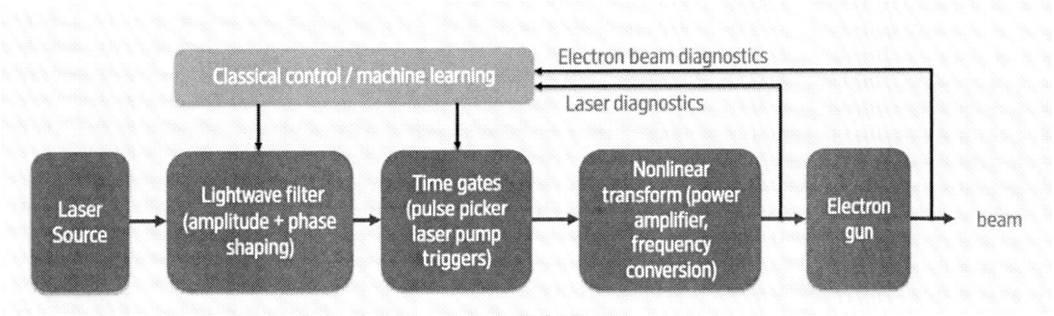

Fig. 1 Flow of the pulse shaping procedure

References

[1] M. Beye, "FLASH2020+: Making FLASH brighter, faster and more flexible : Conceptual Design Report," Verlag Deutsches Elektronen-Synchrotron, Hamburg, 2020. doi: 10.3204/PUBDB-2020-00465.

Deep Learning for Control of Light-Matter Interactions

Ben Mills, Matthew Praeger, Yunhui Xie, Michael McDonnell, Alex Courtier, James Grant-Jacob, and Michalis Zervas
Optoelectronics Research Centre, University of Southampton, UK

The interaction of light and matter is extremely complex, particularly for short laser pulses where nonlinear effects can become the dominant mechanisms. Consequently, even a small, unexpected change in the underlying experimental conditions can result in a severe negative effect on the outcome of the experiment. In practice, the sheer variety and number of experimental parameters that must be monitored means that a human-operator is simply not an effective approach. There is therefore great interest in the development of systems that can provide automated control of laser-based experiments.

In recent years, exciting developments in software and hardware have enabled a revolution in the field of artificial intelligence known as deep learning. Relevant to this presentation, there have been key breakthroughs in the application of deep learning for rapid and precise processing of multiple data streams. An obvious example here is the recent developments in autonomous driving, where multiple neural networks monitor visual data from multiple directions, in real time, and decisions are made on millisecond timescales.

The team at Southampton are applying the technology that supports "self-driving" cars to the control and optimisation of a wide range of light-matter interactions. The analogy here is that the team are solving the scientific and engineering challenges that will enable the "self-driving" lasers of the future that can control and optimise experiments and make critical decisions in real-time, without the need for input by a human operator.

This presentation will provide an overview of recent activity at Southampton in the application of deep learning for a range of light-matter experimental setups, namely:

The real-time application of neural networks in femtosecond laser machining for i) the identification of the laser machining parameters and materials parameters, ii) correcting unexpected beam aberrations, and iii) automatically ceasing laser machining at task completion in the case where the task length is not known prior to machining (see figure for details) [1]. The real-time control of the optical tweezers effect, where reinforcement learning is used to control a laser to translate single particles through a maze [2]. Real-time control via reinforcement learning of the spatial position of laser pulses, via automated control of translation stages, to machine specific and bespoke patterns in a target substrate [3]. The potential for using deep learning for real-time beam shaping in a coherent beam combination arrangement [4].

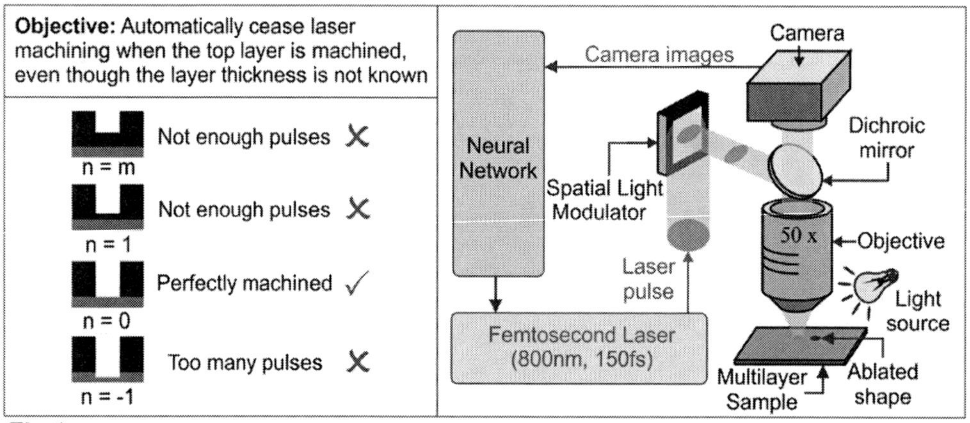

Fig. 1 Example application of deep learning for solving a practical laser machining challenge..

References

[1] Xie, Yunhui, et al. "Deep learning for the monitoring and process control of femtosecond laser machining." Journal of Physics: Photonics 1.3 (2019): 035002.
[2] Praeger, M., Xie, Y., Grant-Jacob, J. A., Eason, R. W., & Mills, B. (2021). Playing optical tweezers with deep reinforcement learning: in virtual, physical and augmented environments. Machine Learning: Science and Technology, 2(3), 035024.
[3] Xie, Y., Praeger, M., Grant-Jacob, J. A., Eason, R. W., & Mills, B. (2022). Motion control for laser machining via reinforcement learning. Optics Express, 30(12), 20963-20979.
[4] Mills, B., Grant-Jacob, J. A., Praeger, M., Eason, R. W., Nilsson, J., & Zervas, M. N. (2022). Single step phase optimisation for coherent beam combination using deep learning. Scientific Reports, 12(1), 1-12.

 https://doi.org/10.1051/epjconf/202226701067

Self-Starting Kerr-Lens-Modelocked 1-GHz Ti:sapphire Oscillator Pumped by a Single Laser Diode

Hanna Ostapenko[1], Toby Mitchell[1], Pablo Castro-Marin[1] and Derryck T. Reid[1]

1. School of Engineering and Physical Sciences, Heriot–Watt University, Edinburgh, EH14 4AS, UK.

Diode pumping of Ti:sapphire [1,2] provides a route to commodifying Ti:sapphire lasers by breaking the reliance on a complex diode-pumped solid-state (DPSS) pump source and, in the context of laser frequency combs, the ability to directly modulate the laser diode current at multi-100-kHz bandwidths provides direct and high-speed control of the carrier-envelope offset frequency [3,4]. With appropriate management of the pump beam quality, soft-aperture Kerr-lens modelocking can be achieved in diode pumped Ti:sapphire lasers, with examples of green [3] and blue [5] pumped lasers being reported with repetition frequencies as high as 420 MHz [3]. The extension to GHz repetition frequencies is made more challenging by the correspondingly lower pulse peak powers available, and so far no example of a diode-pumped GHz Kerr-lens-modelocked Ti:sapphire laser has been reported, to our knowledge. Here, we present an ultra-simple, three-element Kerr-lens-modelocked Ti:sapphire laser. This self-starting oscillator is pumped by a single green laser diode and provides 108-fs pulses with a repetition frequency of 1.01 GHz and an average power >100 mW. Figure 1 presents the laser design and output pulses inferred from the spectrum and autocorrelation measurements.

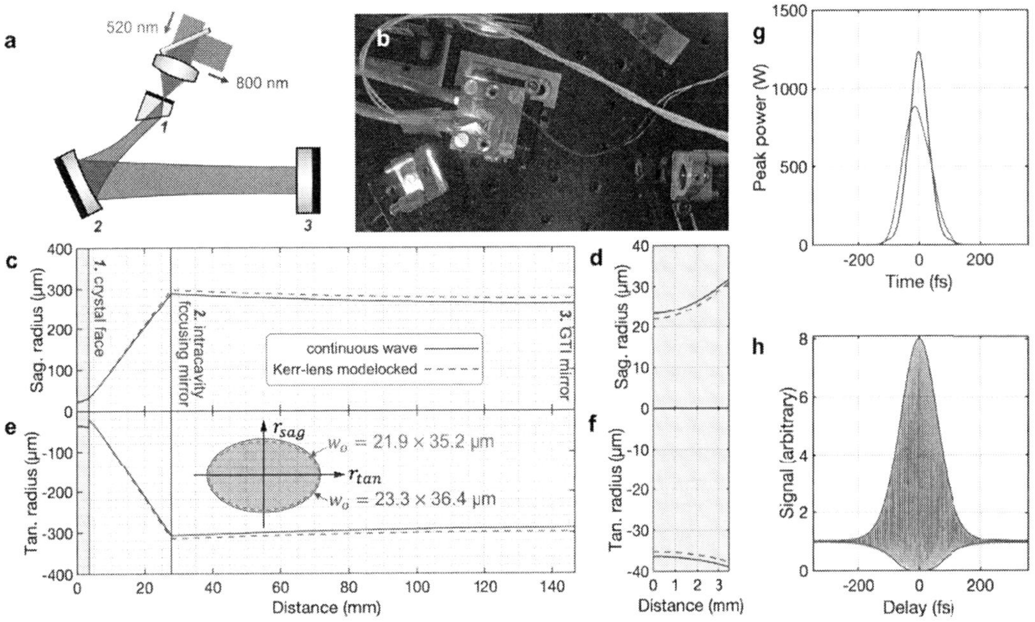

Fig. 1 (a) Layout of the diode-pumped Kerr-lens-modelocked Ti:sapphire laser, and (b) photograph of the operating laser. (c) Sagittal beam radius ($1/e^2$ intensity) throughout the whole cavity and (d) detail of the beam radius inside the Ti:sapphire crystal. (e) and (f) depict the corresponding tangential beam radius. Solid and dashed lines represent the beam profiles in cw and modelocked operation respectively. The inset of (e) shows (to scale) the beam waist profile inside the Ti:sapphire crystal, whose area reduces by ≈10% when modelocking occurs. (g) Transform-limited (blue) and actual (red) pulse shapes inferred from the spectrum and (h) measured interferometric autocorrelation.

This post-deadline talk will present further details of the laser design, along with spatial, temporal and spectral characterisations, RIN measurements and repetition-rate stabilization performance.

References

[1] P. W. Roth, A. J. Maclean, D. Burns, and A. J. Kemp, "Directly diode-laser-pumped Ti:sapphire laser," Opt. Lett. **34**, 3334 (2009).
[2] P. W. Roth, A. J. Maclean, D. Burns, and A. J. Kemp, "Direct diode-laser pumping of a mode-locked Ti:sapphire laser," Opt. Lett. **36**, 304 (2011).
[3] K. Gürel, V. J. Wittwer, M. Hoffmann, C. J. Saraceno, S. Hakobyan, B. Resan, A. Rohrbacher, K. Weingarten, S. Schilt, and T. Südmeyer, "Green-diode-pumped femtosecond Ti:Sapphire laser with up to 450 mW average power," Opt. Express **23**, 30043 (2015).
[4] P. Castro-Marin, T. Mitchell, J. Sun, and D. T. Reid, "Characterization of a carrier-envelope-offset-stabilized blue- and green-diode-pumped Ti:sapphire frequency comb," Opt. Lett. **44**, 5270 (2019).
[5] C. G. Durfee, T. Storz, J. Garlick, S. Hill, J. A. Squier, M. Kirchner, G. Taft, K. Shea, H. Kapteyn, M. Murnane, and S. Backus, "Direct diode-pumped Kerr-lens mode-locked Ti:sapphire laser," Opt. Express **20**, 13677 (2012).

Spectral broadening of 2-mJ ultrashort pulses in a convex-concave multipass cell in ambient air

Alan Omar,* Tim Vogel, Martin Hoffmann, and Clara J. Saraceno
Photonics and Ultrafast Laser Science, Ruhr-Universität Bochum, Universitätsstrasse 150, 44801 Bochum, Germany

Efficient temporal pulse compression of high-average power, ultrafast lasers continues to be a topic of important research efforts. Among many temporal pulse compression techniques suitable for ultrashort pulses with >mJ pulse energy, nonlinear spectral broadening in gas-filled Herriott-type multipass cell (MPC) has been in the spotlight in the last few years, particularly for high average power lasers. Many impressive demonstrations have been realized exhibiting high optical transmission, high compression factors and excellent spatial-spectral homogeneity. For example, in [1], a 1-kW, 1-mJ, 200 fs Yb:fiber laser was compressed from 200 fs to 31 fs with 96% transmission in an argon-filled MPC in a 1.5-m long chamber. In [2], 18 mJ pulses were spectrally broadened from 1.3 ps to a transform-limit of 41 fs in a 3-m long argon-filled MPC. More recently, Pfaff et al. spectrally broadened 840-fs, 10-mJ pulses and showed compressibility to 38 fs also with high optical transmission of 97% with a low-pressure chamber length of 2.7 m [3]. However, all these systems use concave-concave designs in costly pressure controlled vacuum chambers filled with noble gases, and long mirror distances are required to avoid ionization at the focus inside the cell. Here, we demonstrate nonlinear spectral broadening of 2.1-mJ, 670-fs pulses at 210 W of average power in a focus-free compact (61.5 cm distance), convex-concave MPC in ambient air. We show pulse broadening from 21 nm to 24.5 nm, and demonstrate compressibility down to the transform-limit of 134 fs with excellent spectral homogeneity, limited only by damage on the optical mirrors available at the time of the experiment.

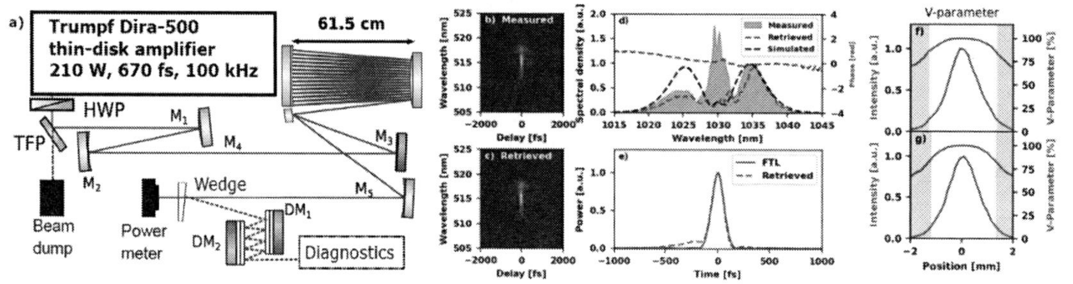

Fig. 1 a) Experimental setup of the thin-disk amplifier, convex-concave MPC, and compression setup. b) and c) measured and retrieved FROG trace. d) Retrieved, measured, simulated spectrum and the spectral phase. e) Retrieved and transform limit of the measured spectrum. f) and g) The red solid curves show the corresponding spatial-spectral homogeneity values (V-parameter).

Fig. 1a) shows the experimental setup. The driving amplifier is a thin-disk regenerative amplifier (DIRA 500-10) operated at 100 kHz of repetition rate, and capable of achieving up to 500 W of average power, with 670 fs pulses. The used MPC consists of two highly-reflective (HR) mirrors, one plano-concave with a radius of curvature (ROC) of 2 m and one plano-convex with a ROC of -2.5 m. The design of this MPC was critical and obtained using a fast 3D pulse propagation code. The main restriction for the input pulse energy was dictated by the available convex mirror at the time of the experiment. With a distance of 61.5 cm, 30 passes (15 roundtrips) in the ambient air were achieved with 96% throughput. Compressibility is shown with a fraction of the pulses (3.5%) using -18100 fs2 of dispersion, and characterized using frequency-resolved optical gating (FROG). Fig. 1b) and c) show the results of the measured and retrieved traces of the compressed pulse with good agreement with each other, with a FROG error of 0.5% using a 512x512 grid. Fig. 1d) shows the measured spectrum (blue-filled) with a good agreement with the retrieved spectrum (dashed red), the simulated spectrum (black dashed) and the retrieved spectral phase (green dashed). The broadened spectrum supports a 133-fs pulse duration. In Fig. 1e), the retrieved temporal intensity profile (dashed) shows a nearly transform-limited pulse (blue) with a pulse duration of 134 fs. In addition, the convex-concave MPC preserves the beam profile quality with $M^2<1.3$ and shows good spectral homogeneity with a V-parameter of both spatial axes measured over 80% in the $1/e^2$ area of the beam profile, which is similar to other gas-filled MPC results, see fig. 1f) and g). The only limitation in this experiment was damage of the optics available at higher power, which can be easily circumvented with other MPC designs and/or higher damage threshold optics. In conclusion, we show nonlinear spectral broadening of 2 mJ pulses in a compact, convex-concave multi-pass cell in ambient air. This system is straightforward to scale to much higher pulse energies in simple and cost-efficient layouts.

References :

1. C. Grebing, M. Müller, J. Buldt, H. Stark, and J. Limpert, "Kilowatt-average-power compression of millijoule pulses in a gas-filled multi-pass cell," Opt. Lett. **45**(22), 6250-6253 (2020).
2. M. Kaumanns, V. Pervak, D. Kormin, V. Leshchenko, A. Kessel, M. Ueffing, Y. Chen, and T. Nubbemeyer, "Multipass spectral broadening of 18 mJ pulses compressible from 13 ps to 41 fs," Opt. Lett. **43**(23), 5877 (2018).
3. Y. Pfaff, C. Forster, G. Barbiero, M. Rampp, S. Klingebiel, J. Brons, C. Y. Teisset, H. Wang, R. Jung, J. Jaksic, A. H. Woldegeorgis, C. J. Saraceno, and T. Metzger, "Nonlinear pulse compression of a thin-disk amplifier and contrast enhancement via nonlinear ellipse rotation," Opt. Express, OE **30**(7), 10981-10990 (2022).

High-power optical amplifier with enhanced wall-plug efficiency for 10-channel WDM satellite laser communication systems

Sven Hochheim[1], Alexander Büttner[1], Eike Brockmüller[1], Willy Fittkau[1], Felix Wellmann[1], Peter Weßels[1], Jörg Neumann[1], Dietmar Kracht[1]

1. Laser Zentrum Hannover e.V., Hollerithallee 8, 30419 Hannover, Germany

As part of the ARTES ScyLight framework programme (SeCure and Laser communication Technology), ESA envisages the establishment of a high throughput optical network in the sky called HydRON [1]. This next generation of optical satellite communication technology foresees the employment of all-optical payloads, targeting for a seamless extension of terrestrial fiber networks with terabit per second capacity.

In this context, we present the development of a fiber-based high-efficiency laser amplifier (HELA). The system is designed to simultaneously amplify 10 narrow-linewidth optical channels (< 15 MHz) at wavelengths around 1064 nm. For input powers ranging between 1 and 10 mW per channel, the targeted output power is up to 10 W per channel (minimum requirement: 5 W), i.e., up to 100 W in total. Considering the strict electrical power restrictions for satellite payloads, a focus was placed on enhancing the wall-plug efficiency of the system.

An in-house developed fiber laser setup combining 10 optical channels (1060 nm to 1069 nm, approx. 1 nm spacing) in a single-mode fiber serves as a seed source. The HELA system is designed as an all-fiber setup which is illustrated in Fig. 1a). It comprises two amplifier stages based on Yb^{3+}-doped polarization maintaining (PM) double-clad fibers. In-house manufacturing technology for fiber optical components enables the integration of pump-combiners (PC) [2], cladding light strippers (CLS) [3] and end caps directly into the active fibers. Therefore, additional splices within the amplifier chain are avoided in favour of preserving the optical properties of the input signal. In between both amplifier sections, two fiber-based optical isolators are employed in order to decouple the booster stages from each other.

With this design, an amplification up to a total output power of 100 W has been demonstrated while employing the aforementioned fiber-based 10-channel seed source. In Fig. 1b) the corresponding amplifier output power is plotted as a function of the current applied to the pump diodes. With respect to the optical power emitted by the two wavelength-stabilized pump diodes of the main amplifier stage, a slope efficiency of around 63% has been achieved. Ongoing development efforts include fiber cut-back experiments in order to optimize the performance of the system. The lengths of the active fibers used in each of the two amplifier stages are a key parameter in defining the amplifier efficiency as well as the achieved gain flatness across the 10-channel seed spectrum.

Considering its intended application in space, irradiation tests have been performed with several components of the HELA system. The in-house manufactured fiber components and the optical isolator have been subjected to gamma irradiation, whereas the employed fiber-coupled pump diodes have been characterized before and after proton irradiation.

While this development is driven by the needs and specifications of the HydRON mission concept, it can also benefit other space applications in the domain of lidar and scientific metrology.

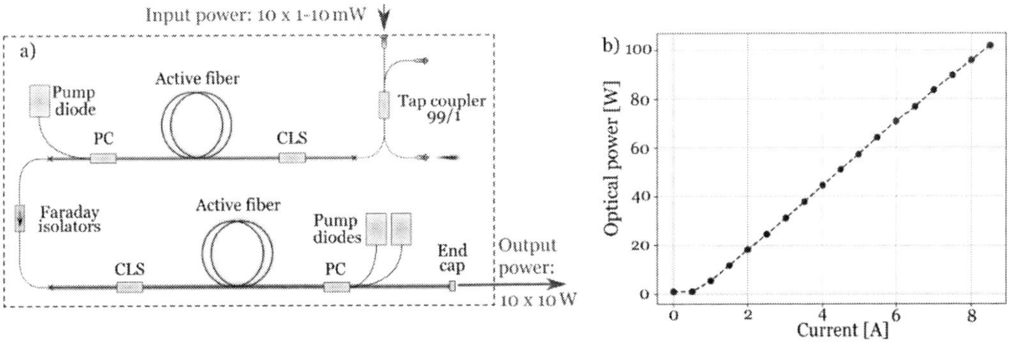

Fig. 1 a) Setup of the HELA system designed as an all-fiber architecture comprising two amplifier stages. PC: pump combiner, CLS: cladding light stripper, x: splice. b) Optical slope of the HELA system up to an output power level of 100 W, plotted as a function of the current applied to the wavelength-stabilized high-power pump diodes.

References

[1] H. Hauschildt et al., "HydRON: High throughput optical network," in Proc. SPIE 10910, Free-Space Laser Communications XXXI, 109100K (4 March 2019). https://doi.org/10.1117/12.2511391

[2] T. Theeg, H. Sayinc, J. Neumann, L. Overmeyer, and D. Kracht, "Pump and signal combiner for bi-directional pumping of all-fiber lasers and amplifiers," Opt. Express **20**(27), 28125-28141 (2012). https://doi.org/10.1364/OE.20.028125

[3] M. Wysmolek, C. Ottenhues, T. Pulzer, T. Theeg, H. Sayinc, M. Steinke, U. Morgner, J. Neumann, and D. Kracht, "Microstructured fiber cladding light stripper for kilowatt-class laser systems," Appl. Opt. **57**(23), 6640-6644 (2018). https://doi.org/10.1364/AO.57.006640

Simple method for determining quantum efficiency and background propagation loss in thulium-doped fibres

Martin P Buckthorpe, William A Clarkson

Optoelectronics Research Centre, University of Southampton, University Road, Southampton, SO17 1BJ, United Kingdom

Thulium (Tm)-doped silica fibre lasers continue to attract much interest as a wavelength flexible and power-scalable source in the two-micron band, benefiting a diverse range of applications from laser surgery and gas sensing to free-space communications and materials processing [1]. One of the key attractions of Tm-doped silica from a power scaling perspective is the 'two-for-one' cross-relaxation process ($^3H_4 + {}^3H_6 \rightarrow {}^3F_4 + {}^3F_4$), which can yield a quantum efficiency approaching '2' when doped with sufficiently high Tm concentrations (>3 wt.%) and pumped at ~793 nm. This opens up the prospect of optical-to-optical efficiencies of >80%. Slope efficiencies of ~70% and higher have been realised in low to moderate power (<100 W) devices [2,3], but scaling to the kilowatt regime has come at the expense of much lower efficiency. One of the key challenges in designing and fabricating Tm fibres for operation at high power levels is maintaining a high Tm-doping level to promote cross-relaxation whilst avoiding the detrimental impact of high thermal loading. This requires more complex core designs to spread heat loading over a longer fibre length, bringing into play the influence of core propagation loss as well as cross-relaxation on overall efficiency. Knowledge of both is essential to understand their relative impact on laser performance, and for further optimising fibre design; however, a direct measurement has not so far been possible.

In this paper, we present a simple method for determining quantum efficiency and background core propagation loss in Tm-doped fibres. The procedure involves setting up a laser utilising a length of Tm fibre with feedback for lasing provided by the Fresnel reflections (~3.5%) from perpendicularly-cleaved facets at both ends of the fibre. Pump light is provided by a fibre-coupled diode laser at 793 nm launched via a free-space arrangement of lenses into one end of the Tm fibre. Laser performance is evaluated by measuring the combined output power from the two ends of the Tm fibre as a function absorbed pump power for different fibre lengths. An example of the results for the Tm-doped nested-ring silica fibre reported in [3] are shown in Fig.1, where slope efficiency with respect to absorbed pump power normalised by the Stokes efficiency, $\eta_s^{abs}\lambda_L/\lambda_P$, is plotted versus fibre length, l.

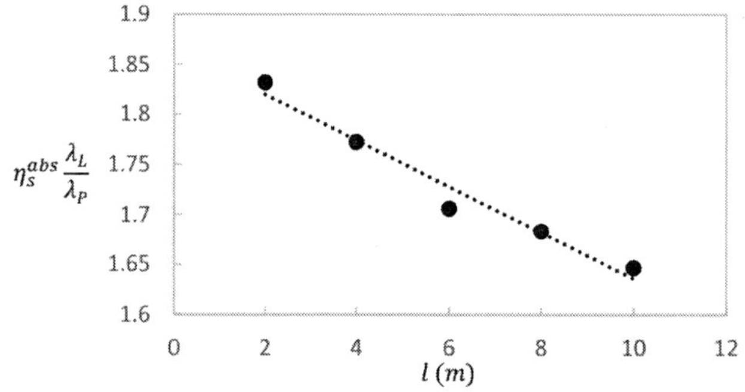

Figure 1: $\eta_s^{abs}\lambda_L/\lambda_P$, versus l for the Tm-doped nested-ring active fibre reported in [3].

An analytical model for laser performance was developed and this shows that the normalised slope efficiency decreases linearly with fibre length, primarily due to core propagation loss. The theory shows that by extrapolating the plot in Fig.1 to the $l = 0$ axis we obtain a value for the quantum efficiency, then from the gradient of a linear fit to the experimental data we can determine the core propagation loss. Details of the model and the operating regimes under which it is valid will be presented. For the Tm-doped nested-ring fibre under study, using the method described above, we determine values for the quantum efficiency and background core propagation loss of 1.87 ± 0.02 and 0.18 ± 0.02 dB/m respectively. It is worth noting that the propagation loss is over an order-of-magnitude higher than would be typically expected for an Yb-doped double-clad fibre. These preliminary findings suggest the key issue to address for power scaling whilst maintaining high efficiency is reducing background loss rather than increasing cross-relaxation efficiency, which is already close to the theoretical limit. The origin of the high background loss and strategies to reduce it are the subject of ongoing studies.

[1] K. Scholle, S. Lamrini, P. Koopmann, and P. Fuhrberg in, *Frontiers in guided wave optics and optoelectronics*, B. Pal ed. (IntechOpen, London,2010), Ch. 22.
[2] P. C. Shardlow et al., "Optimising Tm-Doped Silica Fibres for High Lasing Efficiency", in CLEO/Europe-EQEC '15.
[3] Matthew J. Barber, et al., "Nested-ring doping for highly efficient 1907 nm short-wavelength cladding-pumped thulium fiber lasers." Optics Letters **45.19**, 5542-5545 (2020).

 https://doi.org/10.1051/epjconf/202226702003

Synchronized and tunable femtosecond laser source from CW laser

William Renard[1], Clément Chan[1], Antoine Dubrouil[2], Jérôme Lhermite[3], Giorgio Santarelli[4] and Romain Royon[1]

[1]IRISIOME SAS, 11, Avenue de Canteranne, 33600 PESSAC, France
[2]FEMTOEASY, 11, Avenue de Canteranne, 33600 PESSAC, France
[3]Centre Lasers Intenses et Applications, Université Bordeaux-CNRS-CEA-UMR 5107, 33405 Talence, France
[4]Laboratoire Photonique Numérique et Nanosciences (LP2N), IOGS-CNRS-Université Bordeaux, 33400 Talence, France

Ultrafast fiber-based laser sources are key tools for a large variety of applications in scientific research, medicine, and industry [1]. Fiber oscillators are generally based on mode-locking (ML). Another way to obtain short optical pulses from a CW laser is based on the electro-optic pulse gating approach. This technique allows one to generate pulses down to tens of picoseconds and the repetition rate can be tuned from a few MHz up to a few GHz and synchronization with an external signal with a peak-to-peak timing jitter below 1ps. By using commercially available fibered components and ytterbium-doped fibers, it is straightforward to generate Watt-level optical power around 1.0 μm. With this optical power level and tens of picosecond pulse durations, it is possible to reach the sub-picosecond regime thanks to self-phase modulation (SPM) in long optical fibers. The optical spectrum can be broadened up to a few nm using polarization maintaining (PM) single-mode fiber operating in the normal dispersion regime. Then ultra-short pulses can be retrieved using standard compression techniques. The experimental setup is shown in Figure 1. A 100 mW CW DFB laser diode operating at 1031.6 nm is spliced to a fiber-pigtailed MZM (20 GHz BW) driven by a homemade high-speed pulse generator (45ps FWHM electrical pulses). The pulse repetition rate is controlled by an external RF synthesizer and we operate the laser from 10 MHz to 100 MHz. The intensity modulator is actively bias controlled in order to maximize the Extinction Ratio (ER) between the optical pulses and the CW background (> 30 dB). The optical pulse train obtained at the MZM output is amplified up to 2 W of average power and injected in 150 m of PM980 fiber. Simulations were used as a guide to determine the optimal fiber length in our system, indicating that 150m is sufficient to obtain significant spectral broadening (10nm at -10 dB) [2].

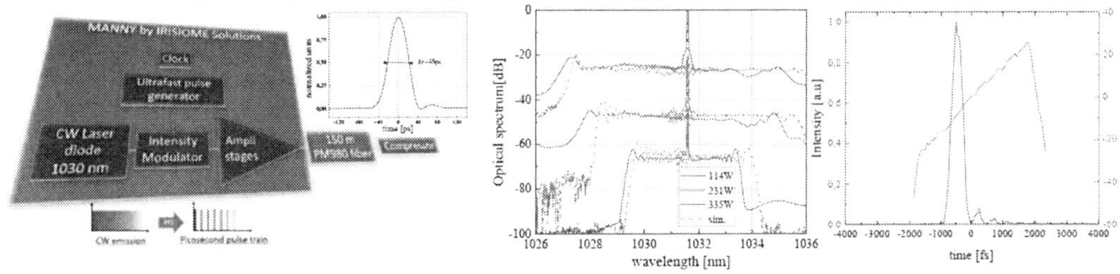

Fig. 1. (Left) Experimental setup for the laser system and measured optical pulse. (Middle) Simulated and measured spectra with different input powers level. (Right) FROG trace of the output pulses and (intensity and phase) at 10MHz repetition rate.

Experimentally, we observe the onset of Stimulated Raman Scattering (SRS) (not considered in the simulations). However, despite the onset of SRS, the spectrum continues to broaden. A saturation of the spectral broadening occurs when the second order Stokes appears [3]. We compressed the output pulses using a homemade grating compressor (1750 l/mm) and measured the compressed pulses (30 nJ) with a single-shot FROG, the FWHM pulse duration is 377 fs (368 fs transform limited pulse). The temporal profile is clean, despite small post-pulses inherited from the ps pulse of the output of the amplifier. Indeed, the initial post pulses has created several post pulses during the nonlinear propagation. However, most of the energy, about 94%, is contained in the main pulse. The retrieved pulse durations are almost the same for the different repetition rates when the injected power is tuned at the input of the nonlinear fiber to obtain same spectral broadening. Usually, the compressor must be reoptimized when changing the laser output power since the chirp and spectral broadening vary with the peak power [4]. Therefore, it is possible to obtain the shortest pulse duration without changing the compressor length, while modifying the pulse repetition rate to realize unprecedented agility. The power tuning is not critical and can support up to 10% of variation with negligible impact on the pulse duration and shape. We have demonstrated an agile and simple alternative to mode-locked lasers. With this experiment we open new avenues in non-mode-locked femtosecond sources for application requiring agility and simple synchronization with low timing jitter. This paves the way to use this technology for instance in micromachining or other applications where high repetition rates are beneficial [5].

References

[1] U. Keller, Appl. Phys. B, **100**, 15-28 (2010).
[2] S. Pierrot and F. Salin, Opt. Express, **21**, 20484-20496 (2013).
[3] A.M. Weiner, J. P. Heritage and R. H. J. Stolen, Opt. Soc. Am. B, **5**, 364-372 (1988).
[4] J.P. Heritage, A.M. Weiner, R.J. Hawkins and O.E. Martinez, Optics Communications, **67**, 367-372 (1988).
[5] W. Renard, C. Chan, A. Dubrouil, J. Lhermite, G. Santarelli and R. Royon, Laser Phys. Lett. **19**, 075105 (2022)

Pulse energy enhancement by means of fiber Bragg gratings in actively Q-switched Tm^{3+}-doped fiber lasers operating at 2050 nm and 2090 nm

Julian Schneider[1,2], Patrick Forster[1,2], Dieter Panitzek[1], Dominik Lorenz[1,2], Clément Romano[1], Marc Eichhorn[1,2], and Christelle Kieleck[1]

[1] *Fraunhofer IOSB (Institute of Optronics, System Technologies and Image Exploitation) Gutleuthausstraße 1, 76275 Ettlingen, Germany*
[2] *Institute of Control Systems, Karlsruhe Institute of Technology, Fritz-Haber-Weg 1, 76131 Karlsruhe, Germany*

Corresponding author: julian.schneider@iosb.fraunhofer.de

Q-switched Tm^{3+}-doped fiber lasers enable the efficient generation of nanosecond pulses at wavelengths around 2 µm in a compact single-oscillator configuration. Their capability of reaching high average output power makes them attractive pump sources for nonlinear frequency conversion into the mid-wave infrared (MWIR), which in turn is ideally suited for spectroscopy, optical countermeasures, and minimally invasive surgery. Methods for frequency conversion into the MWIR include the use of optical-parametric oscillators (OPOs) based on nonlinear crystals such as $ZnGeP_2$ (ZGP) and the generation of a supercontinuum (SC) in fluoride fibers featuring low guiding losses in this spectral range [1,2]. In order to achieve high MWIR output power efficiently, it is necessary to improve the 2 µm pump lasers in terms of pulse energy, peak power, and average output power.

We present a diode-pumped Tm^{3+}-doped actively Q-switched fiber laser optimized for generating high pulse energies by using a fiber Bragg grating (FBG) as an output coupler (OC). Using an FBG OC with a tailored reflectivity at the operating wavelength instead of a straight cleaved fiber end prevents broadband feedback seeding ASE and parasitic lasing, which eventually limit the maximum achievable pulse energy.

Figure 1 shows the pulse energy at constant pump power versus the pulse repetition rate (PRR) of the fiber laser with and without using an FBG at an operating wavelength of 2050 nm. Splicing the FBG on the OC side allows further scaling of the pulse energy by 20 % before parasitic lasing occurs. At the maximum pulse energy of 960 µJ, Gaussian pulses with a pulse width of 44 ns and a peak power of 20.5 kW were obained. The output spectrum shows a narrow FWHM linewidth of < 1 nm and exhibits a SC tail extending beyond 2.4 µm, which contains more than 40 % of the total output power. This highlights the potential of the Q-switched fiber laser to generate a broad high power MWIR SC that may be further extended by pumping a fluoride fiber.

By using another FBG, it was even possible to Q-switch the fiber laser at a wavelength of 2090 nm where ZGP features low absorption losses. The results are shown in Figure 2. The laser reaches a maximum pulse energy of 720 µJ with a pulse width of 105 ns, a peak power of 6.5 kW, and a narrow linewidth < 1 nm.

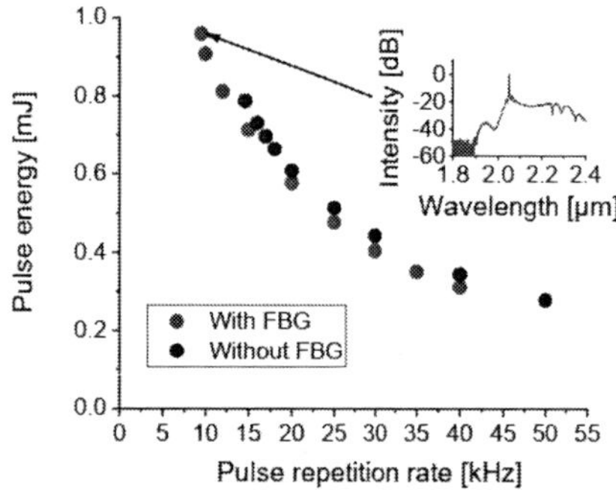

Figure 1: Pulse energy versus PRR of the fiber laser with and without FBG. For the highest pulse energy of 960 µJ a SC appears in the output spectrum.

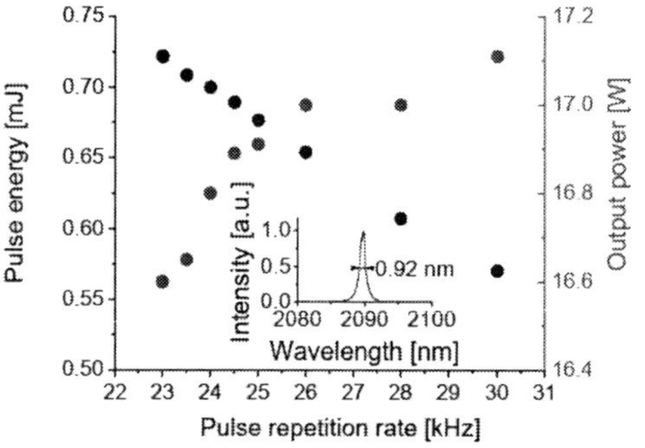

Figure 2: Pulse energy and average output power versus PRR at an operating wavelength of 2090 nm. The inset shows the output spectrum at the highest pulse energy of 720 µJ.

The demonstrated pulse energies and peak powers are unique values for a Q-switched fiber laser operating at wavelengths of 2050 nm and 2090 nm. However, the average output power was < 20 W, limited by pump power. For future work, we plan to significantly increase the average output power by applying two high power pump diodes at 793 nm.

References

[1] J. Schneider, P. Forster, C. Romano, M. Eichhorn, and C. Kieleck, "High pulse energy $ZnGeP_2$ OPO directly pumped by a Q-switched Tm^{3+}-doped single-oscillator fiber laser," Opt. Lett. 46, 2139-2142 (2021).

[2] G. Scurria, I. Manek-Hönninger, J. Carré, A. Hildenbrand-Dhollande, and S. Bigotta, "7 W mid-infrared supercontinuum generation up to 4.7 µm in an indium-fluoride optical fiber pumped by a high-peak power thulium-doped fiber single-oscillator," Opt. Express 28, 7672-7677 (2020).

https://doi.org/10.1051/epjconf/202226702005

Self-generation scheme for heteronuclear compound states

S. Willms[1,2], *S. Bose*[1,3], *O. Melchert*[1,2,4], *U. Morgner*[1,2,4], *I. Babushkin*[1,2], *A. Demircan*[1,2,4]

1. *Cluster of Excellence PhoenixD, Leibniz University Hannover, Welfengarten 1, 30167, Hannover, Germany*
2. *Institute of Quantum Optics, Leibniz University Hannover, Welfengarten 1, 30167, Hannover, Germany*
3. *Institute of Photonics, Leibniz University Hannover, Nienburger Str. 17, 30167, Hannover, Germany*
4. *Hannover Centre for Optical Technologies, Nienburger Str. 17, 30167, Hannover, Germany*

Soliton molecules are usually referred to soliton pairs propagating in waveguides with a fixed time delay, as observed, e.g., in dispersion-managed fibers [1]. Recently, a fundamentally different type of soliton molecule has been proposed [2] and successfully demonstrated in experiments [3]. Such kind of two-color compounds consists of two vastly separated subpulses in the frequency domain with similar group-velocities, enabled by suitable dispersion characteristics with at least two domains of anomalous dispersion (A1,A2) [Fig. 1(a-c)]. We have studied these compound states under perturbations [4] revealing very unique propagation dynamics due to strong entanglement between the individual subpulses. To generate two-color compounds, access to two incommensurable, group-velocity matched frequencies is required, which is difficult to realize experimentally. Here, we propose a self-generation scheme enabled by the Raman effect requiring only a one color pumping. The fiber design is optimized for commercially available wavelengths.

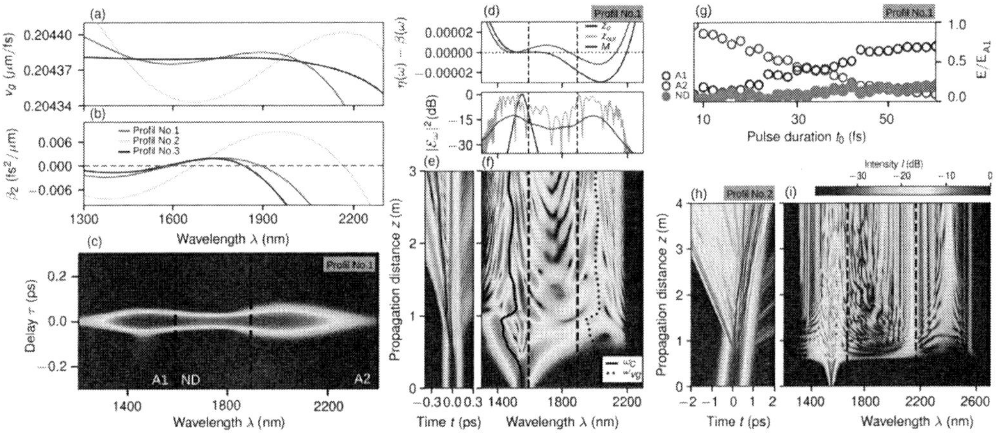

Fig. 1 Self-generation scheme. (a,b) Group-velocity and group-velocity dispersion. (c) Spectrogram of generated compound state. (d) Phase-matching analysis (top) and spectrum (bottom) of input, output, and filtered compound state denoted by dark blue, light blue, and red line, respectively. Propagation dynamics of self-generation process in temporal (e) and spectral (f) domain. (g) Parameter study for initial pulse durations. Supercontinuum generation in temporal (h) and spectral (i) domain.

Suitable group-velocity and group-velocity dispersion profiles are displayed in Fig. 1(a,b), respectively, for three different fiber designs. For the second fiber design a self generation scheme is presented in Fig. 1(d-f). The spectral broadening of an initial higher order soliton accompanied by the Raman-induced frequency down shift [5] initiates the population to a phase-matched component [6][Fig. 1(d); top panel] in A2 [Fig. 1(f)] by the spectral tunneling process [7] . With further propagation this mechanism leads to the build-up of a second group-velocity matched constituent (ω_{gv}) for a stable compound state. Our investigations show that the final energy distribution after a specific propagation distance depends on the initial properties of the soliton as displayed for the example of different pulse durations in Fig. 1(g). Our parameter study shows that we can tailor the process from a complete spectral tunneling to nearly no energy transfer between the regions. This process appears naturally in supercontinuum generation [Fig. 1(h,i)].

References

[1] M. Stratmann, T. Pagel, F. Mitschke, "Experimental Observation of Temporal Soliton Molecules", Phys. Rev. Lett., **95**, 143902 (2005).
[2] O. Melchert, S. Willms, S. Bose et al., "Soliton Molecules with Two Frequencies", Phys. Rev. Lett. **123**, 243905 (2019).
[3] J. P. Lourdesamy, et al., "Spectrally periodic pulses for enhancement of optical nonlinear effects," Nat. Phys. **18**, 59 (2022).
[4] S. Willms, O. Melchert, S. Bose et al., "Heteronuclear soliton molecules with two frequencies," Phys. Rev. A **105**, 053525 (2022).
[5] D. V. Skryabin, F. Luan, J. C. Knight, P. St. J. Russel, "Soliton Self-Frequency Shift Cancellation in Photonic Crystal Fibers", Science, **301**, 1705 (2003).
[6] A. V. Yulin, D. V. Skryabin, and P. St. J. Russell, "Four-wave mixing of linear waves and solitons in fibers with higher-order dispersion," Opt. Lett. **29**, 2411 (2004).
[7] E. N. Tsoy and C. M. D. Sterke, "Theoretical analysis of the self-frequency shift near zero-dispersion points: Soliton spectral tunnelling," Phys. Rev. A **76**, 043804 (2007).

High-peak-power Ho^{3+} and Tm^{3+}-doped fiber MOPA for mid-IR conversion

Dominik Lorenz[1,2], Clément Romano[1], Dieter Panitzek[1], Patrick Forster[1,2], Julian Schneider[1,2], Marc Eichhorn[1,2], and Christelle Kieleck[1]

[1] *Fraunhofer IOSB (Institute of Optronics, System Technologies and Image Exploitation), Gutleuthausstraße 1, 76275 Ettlingen, Germany*
[2] *Institute of Control Systems, Karlsruhe Institute of Technology, Fritz-Haber-Weg 1, 76131 Karlsruhe, Germany*

Corresponding author: dominik.lorenz@iosb.fraunhofer.de

Master-oscillator power amplifiers (MOPA) based on Tm^{3+} and Ho^{3+}-doped silica fibers are ideal pump sources for mid-IR nonlinear frequency conversion as they allow for optimized pump pulse shapes to increase the conversion efficiency in nonlinear converters, e.g. in $ZnGeP_2$ (ZGP) optical-parametric oscillators (OPO) [1]. This allows preserving the beam quality while scaling the average mid-IR output power and makes all-in fiber based MOPAs the most promising candidate for compact and robust OPO pump sources. We present a polarization-maintaining three-stage monolithic fiber MOPA seeded by a laser diode at 2047 nm and its application to pump a ZGP crystal in a linear OPO cavity.

The experimental setup of the MOPA is shown in figure 1. By modulation of the input current a linear pulse shape of the seed laser is created (50 ns pulse width, 50 kHz repetition rate), which allows the incorporation of saturation effects of the power amplifiers while inducing a frequency chirp along the pulse leading to an increased threshold for stimulated Brillouin scattering (SBS). The µW-level seed power is first amplified by a Ho^{3+}-doped-silica fiber, counter pumped by a thulium fiber laser emitting at 1993 nm [2]. Thulium-doped silica fibers are used in the second and third-stage amplifier pumped at 793 nm, which together increase the average signal output power to 19.8 W. By spectral filtering in between the second and third stage spectral degradation caused by modulation instabilities as well as the overall noise level in the output signal are reduced. The pulse peak power is limited by the onset of SBS detected at the backwards port of the tap coupler. A diffraction limited beam quality is obtained and a polarization extinction ratio (PER) >11 dB is measured. After passing an isolator, which induces a reduction of the 2047 nm output power down to 18.4 W, the output of the MOPA is launched into a linear ZGP OPO cavity. In figure 2, the combined signal and idler mid-IR output power is shown versus the launched pump power.

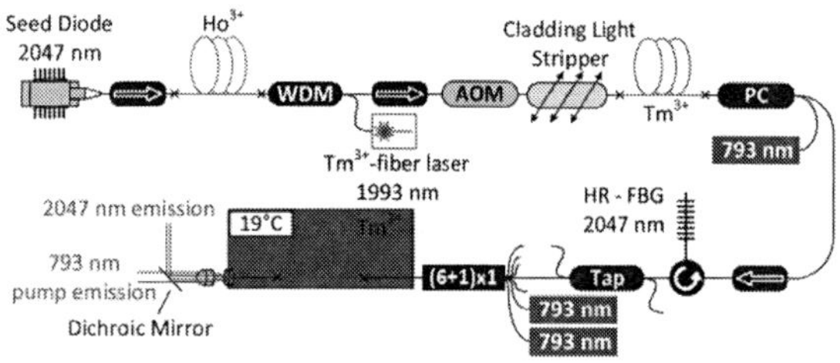

Figure 1: Experimental setup of the all-fiber MOPA.

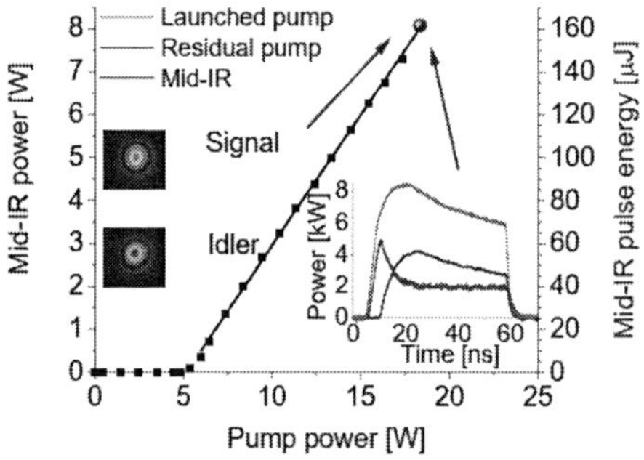

Figure 2: Combined mid-IR power versus launched pump power. The inset on the right shows the launched pump pulse (red) together with the residual pump (green) and the mid-IR pulse (purple). The insets on the left show the beam profile of signal and idler.

A maximum mid-IR output power of 8.1 W (162 µJ pulse energy) is reached at a differential OPO efficiency of 61 % and a total OPO conversion efficiency of 44 %. The launched pump pulse, the residual pump and the combined mid-IR pulse are shown in the right inset of figure 2. The beam quality factors for signal and idler are obtained by separately measuring the beam diameters around the focus spot. An M^2 value of 2.2 and 2.0 is found for signal and idler, respectively. The left insets show the beam profiles of signal and idler.

Further investigations will target the increase of average power and pulse energy by managing nonlinearities in the power stage of the pump laser.

References

[1] Holmen, L. G. and Fonnum, H., "Holmium-doped fiber amplifier for pumping a $ZnGeP_2$ optical parametric oscillator," Opt. Express 29(6), 8477–8489 (2021).

[2] Dominik Lorenz, Clément Romano, Marc Eichhorn, and Christelle Kieleck, "Nanosecond pulsed single-frequency two-stage holmium-doped fiber MOPA at 2054 nm and 2090 nm," Appl. Opt. 60, F27-F32 (2021)

High-power and highly-efficient laser operation of Tm^{3+}:Ho^{3+}-codoped silica fiber lasers emitting at 2.1 μm and 2.2 μm

Patrick Forster[1,2], Clément Romano[1], Julian Schneider[1,2], Marc Eichhorn[1,2], and Christelle Kieleck[1]

[1] *Fraunhofer IOSB (Institute of Optronics, System Technologies and Image Exploitation), Gutleuthausstraße 1, 76275 Ettlingen, Germany*
[2] *Institute of Control Systems (IRS), Karlsruhe Institute of Technology, Fritz-Haber-Weg 1, 76131 Karlsruhe, Germany*

Corresponding author: patrick.forster@iosb-extern.fraunhofer.de

Efficient high-power lasers emitting in the 2.1 μm to 2.2 μm wavelength range are of increasing interest for applications like mid-IR nonlinear frequency conversion or LIDAR, taking advantage of a broad atmospheric transmission window. To emit at wavelengths beyond 2.1 μm, Ho^{3+}-doped fiber lasers (HDFL) are predestined due to their matching emission cross-sections. However, power scaling within this wavelength region turns out to be quite challenging. Inefficient laser operation attributed to pair-induced quenching is leading to complex Tm^{3+}-doped fiber laser (TDFL) pumped setups. Despite a quite promising theoretical quantum defect of less than 10 %, slope efficiencies of this Ho^{3+}-stage are limited to 50-57 % for high laser output powers [1,2].
We present our recent progress in power scaling of alternative Tm^{3+}:Ho^{3+}-codoped silica fiber lasers, especially at emission wavelengths beyond 2.1 μm [3]. We show that high-power laser operation at these wavelengths can be more efficient as in-band pumped high-power Ho^{3+}-fiber lasers with the advantage of a much simpler setup using direct 79X nm diode pumping. Realizing a tunable setup by using a diffraction grating, a 200-nm wavelength span from 1990 nm to 2190 nm is covered with output powers above 10 W. This indicates the high potential for stable high-power laser operation for these wavelengths. Using a volume Bragg grating (VBG) centered at 2.1 μm as a high reflector, stable power scaling at 2.1 μm is established with a maximum output power of 262 W and a slope efficiency of 49 %. The left-hand side of figure 1 shows the according laser output power versus incident pump power.

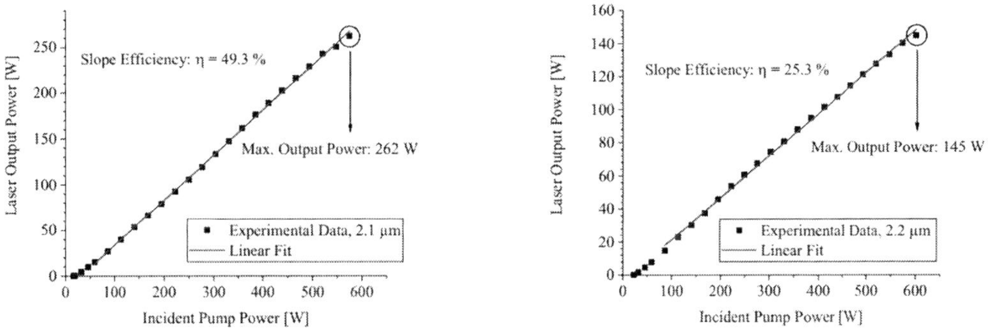

Figure 1: Laser output power versus incident 79X nm pump power of the Tm^{3+}:Ho^{3+}-codoped fiber laser at an emission wavelength of 2.1 μm (left-hand side) and at an emission wavelength of 2.2 μm (right-hand side).

A special challenge is the laser operation at an emission wavelength of 2.2 μm. At this wavelength, the emission cross-sections are vanishingly small for the Ho^{3+}-ions, the silica host shows increased absorption due to the uprising intrinsic IR-absorption and residual OH-contamination in the fiber. Nevertheless, using a VBG centered at 2.2 μm, the Tm^{3+}:Ho^{3+}-codoped fiber laser delivers up to 145 W at 2.2 μm (figure 1, right-hand side), increasing the current power record at this wavelength by more than one order of magnitude [4].

References

[1] A. Hemming, S. Bennetts, N. Simakov, J. Haub, and A. Carter, "Development of resonantly cladding-pumped holmium-doped fibre lasers", Proc. SPIE 8237, 82371J (2012).
[2] A. Hemming, N. Simakov, A. Davidson, S. Bennetts, M. Hughes, N. Carmody, P. Davies, L. Corena, D. Stepanov, J. Haub, R. Swain, and A. Carter, "A monolithic cladding pumped holmium-doped fibre laser", CLEO San Jose, CW1M.1 (2017).
[3] P. Forster, C. Romano, J. Schneider, M. Eichhorn, and C. Kieleck, "High-power continuous-wave Tm^{3+}:Ho^{3+}-codoped fiber laser operation from 2.1 μm to 2.2 μm", Opt. Lett. 47, 2542-2545 (2022).
[4] L. G. Holmen, P. C. Shardlow, P. Barua, J. K. Sahu, N. Simakov, A. Hemming, and W. A. Clarkson, "Tunable holmium-doped fiber laser with multiwatt operation from 2025 nm to 2200 nm", Opt. Lett. 44, 4131-4134 (2019).

Highly efficient side-fused signal pump combiners based on CO_2-laser restructured optical fibers

E. Brockmüller,[1,2] L. Kleihaus,[1] F. Wellmann,[1,2] R. Lachmayer,[3] J. Neumann,[1] D. Kracht[1,2]

1. *Laser Zentrum Hannover e.V., Hollerithallee 8, 30419 Hannover, Germany*
2. *Cluster of Excellence QuantumFrontiers, Welfengarten 1, 30167 Hannover, Germany*
3. *Institut für Produktentwicklung und Gerätebau, An der Universität 1/Gebäude 8143, 30823 Garbsen*

Fiber-based laser and amplifier systems have great potential to provide high output power in combination with high beam quality [1,2] required for a variety of applications ranging from sensor technology and material processing to telecommunications. Such systems achieve high reliability and long-term stability when fiber-based components are used and free-space optics are mostly avoided. Hence, a key fiber component is the signal pump combiner (SPC), which couples the required pump power to the active fiber. The fiber-based side-fused SPC design introduced by Theeg. et al. allows for high pump-power coupling efficiencies while minimizing signal power loss, preserves beam quality, and polarization properties, which enables the amplifier systems to achieve excellent beam properties [3]. It uses tapered coreless intermediate fibers (IFs) to enable efficient coupling of pump power into the target gain fiber. With this design, when using a 25/400-µm LMA-fiber and 200/220-µm pump fibers (PFs), usual pump power to signal fiber cladding coupling efficiency around 90% is experimentally recorded while ray-tracing simulations show a theoretical maximum of 97.8%. As current research focuses on signal output power scaling towards multiple hundreds of watt (which requires multiple hundreds of watt of pump power), even small improvements in coupling efficiency present a significant gain for the overall system performance as well as thermal stability of the fiber components. One significant loss mechanism is the diameter mismatch of the coreless IFs and the core diameter of the PFs. As the pump power from the high-power multimode diodes is delivered by the 200/220-µm PFs, they need to be connected ("spliced") to the IFs to allow for efficient coupling. Because of fiber manufacturing tolerances of the PF and IF diameter, the IF is usually larger compared to the PF core. When simulating a diameter mismatch of 1% the maximum achievable coupling efficiency is reduced by approximately 3%.

In this work, we present the CO_2-laser machining of 200/220-µm pump fibers, where the cladding structure is symmetrically removed over a distance relevant for SPC production. This cladding removal enables the manufacturing of an SPC without splicing the IF and avoiding the fiber diameter mismatch to increase the coupling efficiency. As presented in Fig. 1 a), a coupling efficiency of 92% was experimentally achieved in this

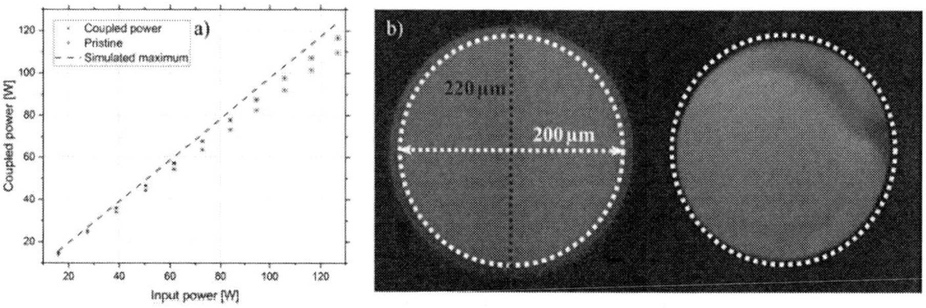

Fig. 1 a) Coupling efficiency over different input powers of an SPC with two pristine pump fibers and an SPC with CO_2-laser machined pump fibers compared to the simulated maximum. b) End facet view of a pristine 200/220-µm pump fiber compared to a CO_2-laser machined sample.

configuration. This presents an increase in coupling efficiency of 4% compared to an SPC manufactured with identical production parameters but a pristine pump fiber. The observed difference to the theoretical maximum is related to manufacturing practical limitations of fiber tapering and fusing. Fig. 1 b) shows the end facets of a pristine pump fiber and a restructured fiber where the entire outer cladding was symmetrically ablated. This work presents an important step towards further power scaling in fiber-based amplifier systems by improving efficiency and high-power reliability. The process could also be transferred to different fiber diameters and other fiber types.

References

[1] S. Hochheim, A. Büttner, P. Weßels, E. Brockmüller, W. Fittkau, J. Neumann, D. Kracht, "100W optical amplifier for 10 channel laser communication system with enhanced wall-plug efficiency in the 1µm wavelength range," Proc. SPIE 11993, 106–111 (2022).

[2] F. Wellmann, M. Steinke, F. Meylahn, N. Bode, B. Willke, L. Overmeyer, J. Neumann, and D. Kracht, "High power, single-frequency, monolithic fiber amplifier for the next generation of gravitational wave detectors," Opt. Express 27, 28523–28533 (2019).

[3] T. Theeg, H. Sayinc, J. Neumann, L. Overmeyer, and D. Kracht, "Pump and signal combiner for bi-directional pumping of all-fiber lasers and amplifiers," Opt. Express 20, 28125–41 (2012).

Synchronized all-PM-fiber Yb-doped amplifiers for high power fs- and ps-pulse generation

Philipp König [1], Jean-Paul Yehouessi [2], Alexandre Gognau [2], Simon Boivinet [2], Antonio Baylon [3], Jean-Bernard Lecourt [2], Yves Hernandez [2], Andreas Wienke [1,4], Uwe Morgner [1,4,5], Jörg Neumann [1,4], Dietmar Kracht [1,4]

1. Laser Zentrum Hannover, Hollerithallee 8, 30419 Hannover, Germany
2. Multitel Innovation center, rue Pierre et Marie Curie, 2, 7000 Mons, Belgium
3. Euro-Multitel, rue Pierre et Marie Curie, 2, 7000 Mons, Belgium
4. Cluster of Excellence PhoenixD (Photonics, Optics, and Engineering Innovation Across Disciplines), Hannover, Germany
5. Institut für Quantenoptik, Leibniz Universität Hannover, Welfengarten 1, 30167 Hannover, Germany

In life science, a combination of several imaging methods can lead to a more detailed view of cancerous cells. Current laser based microscopes usually acquire the image with only one microscopy technique or need a large, complex and expensive laser system to achieve the necessary laser parameters for multiple imaging approaches. In the CARMEN research project, we aim to develop a compact multimodal laser-based microscope that combines Anti-Stokes Raman Spectroscopy as well as multi-photon and stimulated emission depletion (STED) microscopy. For these microscopy techniques, overlapping of two pulses with durations in the femtosecond (fs) and picosecond (ps) range and tunable wavelengths are desirable. These laser pulses can be generated by two tunable optical parametric oscillators (OPOs) to cover the visible spectral range.

Here we present a two-arm, all-PM-fiber Ytterbium amplifier system with femtosecond and picosecond pulse durations for pumping these OPOs. An overview of the laser system is depicted in Figure 1. To synchronize the amplifiers, they are seeded by a single mode-locked fiber oscillator with a repetition rate of 78.9 MHz at a central wavelength of 1064 nm and a pulse duration of 1.5 ps. Afterwards the seed signal is split towards the two separated amplifier arms.

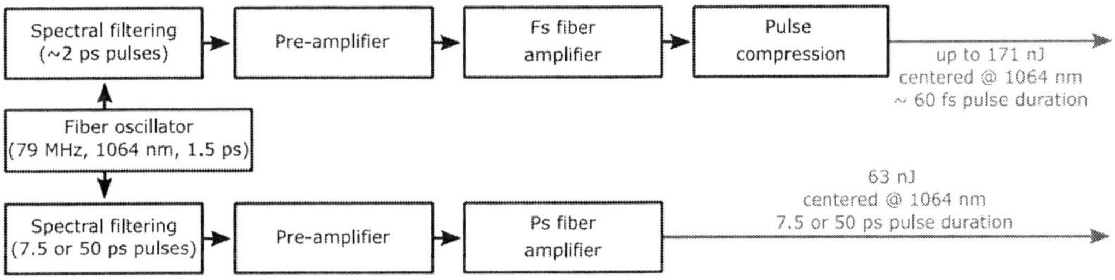

Fig. 1 Schematic overview of the laser system with its two amplifer arms.

In the picosecond arm, the seed pulses are spectrally narrowed by selecting between two fiber Bragg gratings (FBG), controlled by an optical switch. After a pre-amplification stage the pulses are amplified in an Ytterbium-doped large mode area fiber with a 15 μm core to an average power of 5 W (63 nJ pulse energy). Depending on the FBG reflection bandwidth, the Fourier-limited pulse durations are 7.5 ps or 50 ps with corresponding spectral widths of 0.5 nm or 0.05 nm, respectively.

The femtosecond arm is based on nonlinear gain-managed amplification [1]. For improved output pulse characteristics, a band-pass filter and a pre-amplifier ensure a Gaussian shaped spectrum and sufficient seed pulse energies. The main amplification stage consists of a 2 m Ytterbium-doped double-cladding fiber with a 10 μm core. To achieve the desired pulse duration below 100 fs, spectral broadening of the 7 nm wide input spectrum is necessary. This is realized via self-phase modulation in the active fiber, which broadens the spectrum to a width of about 60 nm. The maximum power at the fiber output is 13.5 W, which corresponds to a pulse energy of 171 nJ. The pulses are dechirped by a free space grating compressor to durations below 60 fs.

References

[1] Sidorenko, Pavel, Walter Fu, and Frank Wise. "Nonlinear ultrafast fiber amplifiers beyond the gain-narrowing limit." Optica **6.10** 1328-1333 (2019)

S^2-Method-Based Monitoring of Modal Composition in Optical Fibers during Fiber Component Manufacturing

N. Haverland,[1] *F. Wellmann,*[1,2] *J. Neumann,*[1,2] *D. Kracht*[1,2]

1. Laser Zentrum Hannover e.V., Hollerithallee 8, 30419 Hannover, Germany

2. Cluster of Excellence QuantumFrontiers, Welfengarten 1, 30167 Hannover, Germany

Fiber-based laser systems provide high output power, excellent beam quality and allow easy handling due to an all fiber-based setup and are therefore of high interest for several industries and applications, including material processing and state-of-the-art research. The reliability and stability of fiber-based laser systems is strongly connected to the use of fiber-based components (e.g. signal-pump combiners, cladding light strippers and fiber end caps). These fiber-based systems are often based on large mode-area fibers (LMA), which inherently support higher order modes (HOMs) and thus do not provide a pure single-mode output beam [1]. Recent research activities have developed several new fiber characterization methods in order to analyze the modal content of a beam emitted from a fiber: One example is the S^2-method [1]. The S^2-method was first introduced by Nicholson et al. and was later improved upon by Gray et al., who used a tunable laser source and a CMOS-camera, enabling data acquisition in under a second, making it possible for the S^2-method to be used in real-time applications [1, 2]. In this work we use an in-house developed high-speed S^2-method-based instrument to evaluate the modal content of the emitted beam out of a fiber, which is used for manufacturing of fiber components. This approach has, to the best of our knowledge, not been demonstrated so far. The application of the instrument enables improved manufacturing processes to obtain the best possible performance of the fiber components and thus the fiber laser setup. Specifically, the measurements shown in Figure 1 were carried out during the process of splicing a fused silica endcap to the end of an optical fiber (20 μm core with NA: 0.08, 400 μm cladding diameter) as needed for high-power laser development [3].

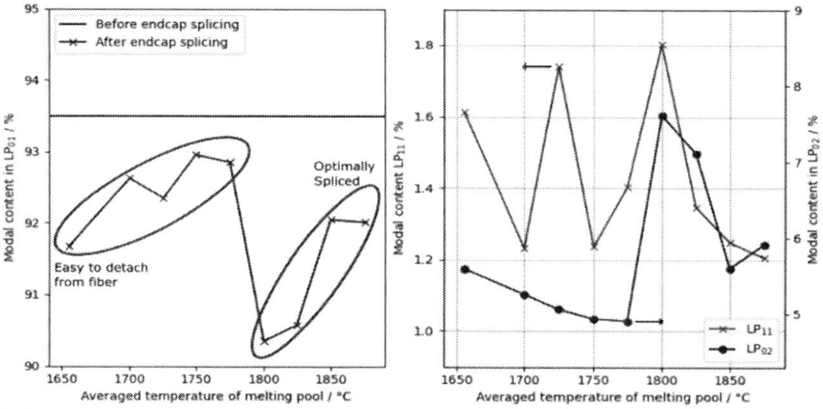

Fig. 1 a) Modal content of LP_{01}-mode compared to modal share before splicing; b) modal content of LP_{11} and LP_{02}

The instrument operates as follow: By synchronizing a fiber-based Fabry-Perot filter, which is used to scan a wavelength range from 1015 nm to 1040 nm, with a CMOS-camera, we were able to obtain up to 400 images per second, with a repetition rate of 0.33 Hz, leading to 30 S^2-method based measurements of the modal composition of the beam during the cooling phase of the endcap-splicing process. We used this method, to optimize the endcap splicing process, by varying the averaged melting pool temperature and observing their influence on the modal content. Figure 1 shows the modal share of the fundamental mode LP_{01} (Fig. 1a) and the higher modes LP_{11} and LP_{02} (Fig. 1b). Even though the fundamental mode content increases with the splicing temperature up to 1775 °C, no sufficient splice between the substrate and the fiber was observed. Above 1800 °C stable splices were achieved, but a decreased splice LP_{01} mode was observed. This is likely related to an insufficient melt pool temperature that cause fiber deformation. With increased melt pool temperatures the LP_{01} content increases, indicating improved splicing conditions, which lead to lower defomation of the optical fiber.

References

[1] J. W. Nicholson, A. D. Yablon, S. Ramachandran, and S. Galmi, "Spatially and spectrally resolved imaging of modal content in large-mode-area fibers," Optics Express, Vol. **16**, No. 10 (2008), 7233-7243.

[2] D. R. Gray, M. N. Petrovich, S. R. Sandoghchi, N. V. Wheeler, N. K. Baddela, G. T. Jasion, T. Bradley, D. J. Richardson, and F. Poletti, "Real-Time Modal Analysis via Wavelength-Swept Spatial and Spectral (S^2) Imaging," IEEE Photonics Technology Letters **28**, No. 9 (2016), 1034-1037.

[3] F. Wellmann, M. Steinke, F. Meylahn, N. Bode, B. Wilke, L. Overmeryer, J. Neumann, and D. Kracht, "High power, single-frequency, monolithic fiber amplifier for the next generation of gravitational wave detectors," Optics Express **27**, No. 20 (2019), 28523-28533.

EPJ Web of Conferences **267**, 02011 (2022)
EUROPHOTON 2022
https://doi.org/10.1051/epjconf/202226702011

Versatile GHz Burst-Mode Operation in High-Power Femtosecond Laser

Tadas Bartulevičius[1], Mykolas Lipnickas[1], Karolis Madeikis[1], Raimundas Burokas[1,2] and Andrejus Michailovas[1,2]

1. Ekspla uab, Savanoriu ave. 237, LT-02300 Vilnius, Lithuania

2. Center for Physical Sciences and Technology, Savanoriu ave. 231, LT-02300, Vilnius, Lithuania

Ultrashort pulse lasers are highly applicable in science, medicine, and industry. The recent development of new techniques which use gigahertz (GHz) bursts of ultrashort pulses attracted a lot of attention by demonstrating the superior performance of laser systems and the advantages in their applications [1]. Common methods for producing GHz bursts of pulses may be implemented in commercial lasers, however, only a few of them can provide significant improvements with a unique set of features and parameters. GHz burst formation by pulse repetition rate (PRR) multiplication techniques based on splitting and delaying ultrafast radiation from the MHz master oscillator are a convenient solution successfully demonstrated experimentally. Moreover, these techniques are of a high potential to be easily implemented in the all-in-fiber design, to provide compact, stable, and robust ultrafast lasers for the science and industry.

A new versatile patent-pending method to generate ultra-high (>2 GHz) repetition rate bursts of ultrashort laser pulses, overcoming many limitations encountered by other fiber-based techniques, is introduced in this work [2]. The developed method was based on the use of an all-in-fiber active fiber loop (Fig. 1 (left)) which contains the main parts responsible for burst formation (acousto-optic modulator (AOM)), pulse amplification (ytterbium-doped fiber (YDF) pump by a laser diode (LD)), and dispersion compensation (chirped fiber Bragg grating (CFBG)). The active fiber loop can form bursts of laser pulses with any desired PRR which does not depend on the initial PRR of a fiber oscillator. Moreover, the bursts of ultrashort laser pulses containing any number of pulses in a burst (from 2 pulses to thousands of pulses inside the burst) with identical intra-burst pulse separation were demonstrated (Fig. 1 (right)). Long bursts even up to a 1 μs width were formed using this method along with a temporal modulation of the initial pulse train. The amplitude of pulses within a burst was controlled by amplification conditions in the YDF amplifier providing a constant, decaying, or rising amplitude of the pulses within a burst as long as it was not limited by the effect of the gain saturation. The burst shape can be additionally modified using an AOM controlled by an arbitrary waveform generator outside the active fiber loop. The dispersion compensation mechanism using the CFBG element was successfully implemented in the active fiber loop which allowed to obtain ultrashort pulse durations within the GHz bursts.

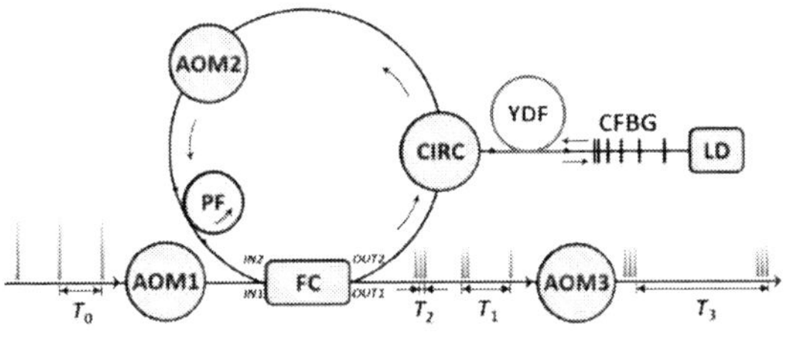

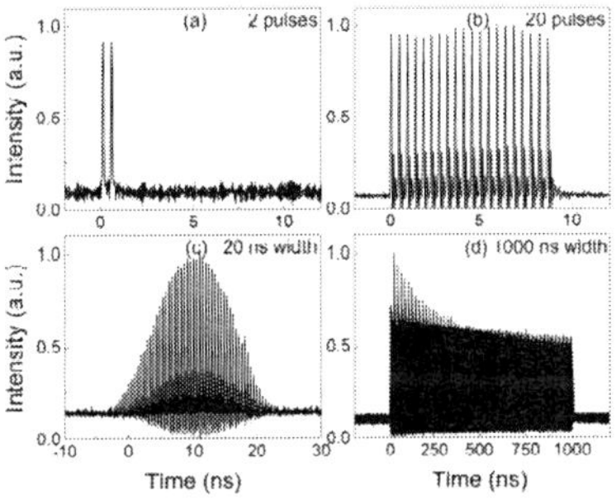

Fig. 1 Left: Schematic representation of an active fiber loop. Time delays: T_0 – between single input pulses, T_1 – between a delayed replica of an input pulse and an undelayed replica of the pulse, T_2 – intra-burst pulse separation of the formed bursts, and T_3 – between bursts of pulses. Right: Experimentally measured 2.2 GHz intra-burst PRR bursts containing from 2 (a) to about 2200 (d) ultrashort pulses.

The active fiber loop was installed into the constructed 30 W-level average power ultrashort (sub-1 ps) pulse laser system which operated in single-pulse and GHz-burst (2.2 GHz intra-burst PRR) operation regimes. A detailed description of a new pulse multiplication technique and the demonstration of its versatility in industrial-grade high-power laser will be provided at the conference.

References

[1] C. Kerse, H. Kalaycıoğlu, P. Elahi, B. Çetin, D. K. Kesim, Ö. Akçaalan, S. Yavaş, M. D. Aşlk, B. Öktem, H. Hoogland, R. Holzwarth, and F. Ö. Ilday, "Ablation-cooled material removal with ultrafast bursts of pulses," Nature **537**, 84–88 (2016).
[2] T. Bartulevicius, K. Madeikis, L. Veselis, V. Petrauskiene, and A. Michailovas, "Active fiber loop for synthesizing GHz bursts of equidistant ultrashort pulses," Opt. Express **28**, 13059-13067 (2020).

EPJ Web of Conferences **267**, 02012 (2022) https://doi.org/10.1051/epjconf/202226702012
EUROPHOTON 2022

Fiber-based light source with multi-color output and fast wavelength tuning

Kristin Wallmeier[1], Thomas Würthwein[1], Maximilian Brinkmann[2], Tim Hellwig[2], Carsten Fallnich[1,3]
1. University of Münster, Institute of Applied Physics, Corrensstraße 2, 48149 Münster, Germany
2. Refined Laser Systems GmbH, Mendelstraße 11, 48149 Münster, Germany
3. Cells in Motion Interfaculty Centre, University of Münster, Münster, Germany

Fiber lasers are of high interest for a variety of applications due to their robustness and small footprint. In combination with a synchronized output at different wavelengths, they provide a powerful tool, e.g. for the field of microscopy and spectroscopy. Therefore, we present a fiber-based light source with fast tunable multi-color output and low-noise characteristics.

The light source (schematic in Fig. 1(a)) consists of an oscillator and a subsequent fiber-based optical parametric oscillator (FOPO) [1], enabling synchronized output of pulses at two wavelengths with a repetition rate of 40 MHz. An electronically tunable filter allows to tune the oscillator wavelength between 1020 nm and 1060 nm within 5 ms per arbitrary wavelength step.

By exploiting the parametric four-wave mixing process in a photonic crystal fiber, frequency-shifted light is generated in the FOPO. Due to a steep phase matching curve of the FOPO the relatively narrow tuning of the oscillator wavelength leads to a broad tuning of the FOPO wavelength from 750 nm to 950 nm. In addition to the fast wavelength-switching mechanism of the oscillator, a frequency modulation (FM) [2] scheme was implemented in the FOPO to enable pulse-to-pulse wavelength-switching of the FOPO pulses, which was verified by the measurements shown in Fig. 1(b) and (c). The FM scheme allows for probing different signals almost simultaneously, and therefore, can be used for real-time background correction in microscopy purposes [2].

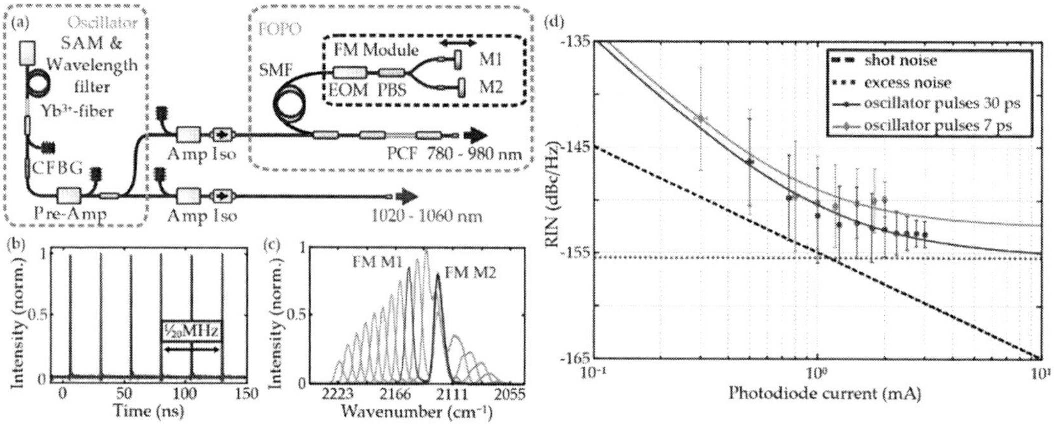

Fig. 1 (a) Schematic of the fiber-based light source with saturable absorber mirror (SAM), chirped fiber Bragg grating (CFBG), amplifiers (Amp), isolators (Iso), single-mode fiber (SMF), photonic crystal fiber (PCF), frequency modulation (FM) unit and mirrors (M). (b) Pulse-to-pulse wavelength-switching by the FM unit at half of the repetition rate of the oscillator. (c) Wavelength tunability within the four-wave mixing gain. (d) Relative intensity noise (RIN) for different pulse durations at variable photodiode currents.

As fiber lasers typically show high noise levels, that are not compatible with sensitive microscopy applications, the relative intensity noise (RIN) was measured. Different parameters like the pulse duration can effect the RIN [3], hence this aspect was investigated in more detail. By increasing the pulse duration from 7 ps to 30 ps due to spectral filtering, the RIN of the amplified oscillator pulses decreased by approx. 3 dB to -153.7 dBc/Hz [4], limited by excess noise around -155.5 dBc/Hz [5] as shown in Fig. 1(d).

In conclusion, the presented fiber-based light source is well suited, e.g. for microscopy or spectroscopy applications, due to its tunable multi-color output and low-noise operation.

References
[1] M. Brinkmann, A. Fast, T. Hellwig, I. Pence, C.L. Evans, and C. Fallnich, "Portable all-fiber dual-output widely tunable light source for coherent Raman imaging", Biomed. Opt. Express **10**, 4437 (2019).
[2] T. Würthwein, M. Brinkmann, T. Hellwig, K. Wallmeier, and C. Fallnich, "High-sensitivity frequency modulation CARS with a compact and fast tunable fiber-based light source", Opt. Lett. **46**, 3544 (2021).
[3] R. Paschotta, "Noise of mode-locked lasers (Part II): Timing jitter and other fluctuations", Appl. Phys. B 79, 153 (2004).
[4] T. Würthwein, K. Wallmeier, M. Brinkmann, T. Hellwig, N.M. Lüpken, N.S. Lemberger, and C. Fallnich, "Multi-color stimulated Raman scattering with a frame-to-frame wavelength-tunable fiber-based light source", Biomed. Opt. Express **12**, 6288 (2021).
[5] H. Yoshimi, K. Sumimura, and Y. Ozeki, "An Er fiber laser generating multi-milliwatt picosecond pulses with ultralow intensity noise", Jpn. J. Appl. Phys. **57**, 108001 (2018).

EPJ Web of Conferences **267**, 02013 (2022)
EUROPHOTON 2022 https://doi.org/10.1051/epjconf/202226702013

1875-nm high-energy mode-locked thulium fiber laser

Panuwat Srisamran, Duanyang Xu, Ibrahim Abughazaleh, Matthew Gerard, Sijing Liang, David Richardson and Lin Xu

Optoelectronics Research Centre, University of Southampton, Southampton SO17 1BJ, United Kingdom

The reduced scattering at long wavelengths and transparency window of typical biological tissues means there is great interest in developing practical high energy femtosecond lasers operating in the short-wavelength infrared (SWIR) region to help increase imaging depth in nonlinear optical microscopy. The wavelength range 1800-1900 nm is of particular interest since it sits between the strong absorption peaks associated with water and lipids - two of the main components of biomedical samples [1]. Thulium (Tm) doped fiber (TDF) has an emission bandwidth that extends from ~1650 nm to 2050 nm and can be used as a gain medium for femtosecond fibre lasers. Mode locked thulium doped fiber lasers (ML-TDFLs) have been demonstrated at the short wavelength end of this emission bandwidth (e.g. at 1740 nm, albeit with a low pulse energy of 0.2 nJ [2]). Conversely, high pulse energy operation has been demonstrated at wavelengths >1900 nm (e.g. 8 nJ at 1920 nm, albeit in a cavity incorporating bulk-optic components [3]). However, the combination of operation at wavelengths in the range 1800-1900 nm and high pulse energy is challenging due to the quasi 3-level nature of TDF and the need for careful management of intracavity nonlinearity, dispersion and filtering respectively. Here we demonstrate, for the first time, a fully-fiberized ML-TDFL laser operating at 1875 nm capable of >10 nJ pulse energies.

A schematic of our mode-locked fiber laser cavity is shown in Figure 1 (a), in which a 0.65-m-length of in-house fabricated TDF with a core diameter of 8 μm and high pump absorption (~20 dB/m at 1580 nm) was used as the gain fiber [4]. The short length of TDF provides efficient gain in the SWIR region and reduces nonlinearity enabling high pulse energy operation. The TDF was pumped by an erbium doped fiber amplifier (EDFA) at 1565 nm through a wavelength division multiplexer (WDM). A semiconductor saturable absorber mirror (SESAM, SAM-1920-36-10ps from Batop Ltd) with a modulation depth of 20% was employed to realize self-start mode locking. A 5-m-length of dispersion compensating fiber (DCF, UHNA4 from Coherent Ltd) was used to manage the intra-cavity dispersion. All the fibers and components in the cavity are non-polarization maintaining (PM) except the isolator. A polarization controller (PC) was placed in front of the PM isolator to form an effective Lyot filter and thereby the filtering needed to support dissipative-soliton operation at SWIR wavelengths. The total cavity length was 11.6 m and the estimated net cavity dispersion was 0.07 ps^2.

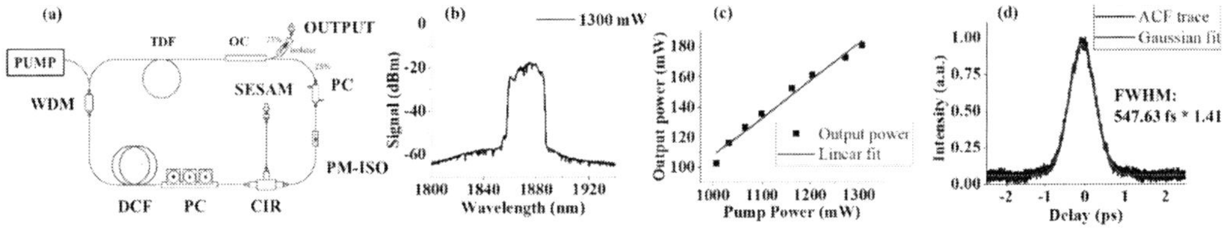

Figure 1 (a) Schematic of the laser, (b) Output spectrum, (c) Output power, and (d) Compressed pulse duration of the laser

The onset of stable single-pulse mode locking was observed at a pump powers of 1 W. The output spectrum of the pulses is shown in Figure 1 (b), which had a central wavelength of 1875 nm and a 3-dB bandwidth of 16 nm. The output power increased with pump power and reached a maximum of 180.5 mW at a pump power of 1.3 W, as shown in Figure 1 (c). At powers beyond this the mode locking became unstable. The measured pulse repetition rate was 17.5 MHz, consistent with the cavity length. The corresponding maximum output pulse energy was 10.3 nJ. The pulse duration at the output port was measured to be 12.44 ps and the pulses could be compressed down to 547.6 fs using a pair of transmission gratings with 900 lines/mm groove density, as shown in Figure 1 (d).

In conclusion, we have demonstrated a high-energy mode-locked Tm-doped fiber laser operating at 1875 nm. The laser generated pulses with a maximum pulse energy of 10.3 nJ which could be compressed extra-cavity to a duration of 547.6 fs. Our SWIR fibre laser represents a promising source for various biomedical applications.

References

[1] R.H. Wilson, K.P. Nadeau, F.B. Jaworski, B.J. Tromberg, and A.J. Durkin, "Review of short-wave infrared spectroscopy and imaging methods for biological tissue characterization", J. Biomed. Opt. **20(3)**, 030901 (2015).

[2] J.-X. Chen, X.-Y. Li, T.-J. Li, Z.-Y. Zhan, M. Liu, C. Li, A.-P. Luo, P. Zhou, K.K.Y. Wong, W.-C. Xu, and Z.-C. Luo, "1.7-μm dissipative soliton Tm-doped fiber laser", Photon. Res. **9(5)**, 873-878 (2021).

[3] Y. Tang, A. Chong, and F.W. Wise, "Generation of 8 nJ pulses from a normal-dispersion thulium fiber laser", Opt. Lett. **40(10)**, 2361-2364 (2015).

[4] M.D. Burns, P.C. Shardlow, P. Barua, T.L. Jefferson-Brain, J.K. Sahu, and W.A. Clarkson, "47 W continuous-wave 1726 nm thulium fiber laser core-pumped by an erbium fiber laser", Opt. Lett. **44(21)** 5230-5233 (2019).

Rapid characterisation of Photonic Crystal Fibre dispersive properties by a stochastic and tunable picosecond pump source

Guillaume Walter[1], Sidi-Ely Ahmedou[2], Thelma De Thoury[1], Nicolas Dos Santos, Jules Herbuvaux[1], Melvin Redon[1], Jean-Christophe Delagnes[1], Romain Dauliat[2], Sébastien Février[2], Constance Valentin[1], Stéphane Petit[1], Nicolas Valero[1], Denis Marion[1], Jérôme Lhermite[1], , Sébastien Tanzilli[4], Frédéric Gérôme[2], Benoît Debord[2], Fetah Benabid[2], Guy Millot[3], Philippe Roy[2], Raphael Jamier[2]

1. *CELIA, Centre Lasers Intenses et Applications, Université de Bordeaux-CNRS-CEA, UMR 5107, F-33405 Talence Cedex, France*
2. *Université de Limoges, CNRS, XLIM, UMR 7252, F-87000 Limoges, France*
3. *ICB, Université de Bourgogne Franche-Comté, CNRS, UMR 6303, F-21078 Dijon, France*
4. *Univ. Côte d'Azur, CNRS, Institut de Physique de Nice, F-06108 Nice Cedex 2, France*

Photonic crystal fibre (PCF) are used in telecom, fibre lasers or nonlinear sources. Due to the silica20 nonlinearity, PCF support Raman conversion, four wave mixing (FWM) generation or continuum generation [1]. The dispersion properties governing these nonlinear effects strongly depend on the PCF transverse section profile. Slight changes in the geometry of the transverse section profile may occur during the fibre drawing process and may alter the generated spectra. It is thus crucial to characterise the dispersion and its higher order accurately and rapidly. Compared to conventional white light interferometry, FWM generation gives access to a wide range of wavelengths with a greater accuracy especially on the zero dispersion wavelength and other PCF characteristics [2]. Within this context, we propose a high-power and widely tunable source based on spectrally shaped amplified spontaneous emission (ASE) with a variable spectral width. Its unique characteristics lead to the observation of FWM with a greatly reduced threshold, and its tunability allows the reconstruction of the dispersion curve from the visible to the mid-infrared (MIR) [2]. The source is depicted in Fig.1(a). It consists in an Ytterbium-doped ASE seeder, emitting a spectrum from 1020 to 1070 nm. A programmable spectral filter is used to select a 0.1-1 nm bandwidth. After a pre-amplification stage, the light is injected in a main amplifier that also modulates the signal periodically thus delivering 50 ps bursts of ASE at 10 MHz. We can generate up to a maximum energy of 2 μJ per 50 ps burst for the PCF under study. At the PCF output, the pump, the signal and the idler waves resulting from the FWM are separated using dichroic mirrors and analysed in terms of spectrum, power and shot-to-shot statistics. Due to the stochastic nature of ASE bursts, a drastic increase of nonlinear effects is expected and observed [3]. The FWM threshold can be decreased by several orders of magnitude as compared to a non-stochastic source obtained from a CW laser (Fig.1(b)).

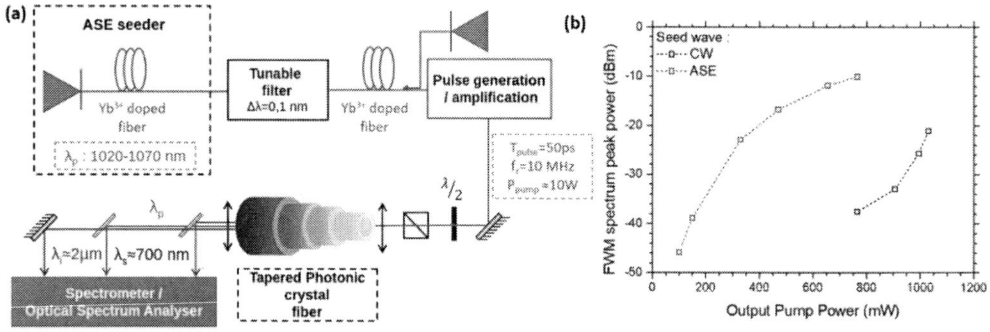

Fig. 1 (a) Experimental scheme of the PCF characterisation source. (b) Evolution of the signal power generated though FWM as a function of the pump power for an ASE seed or a Continuous Wave seed.

We used this source to characterise PCFs and tapered PCFs optimised for MIR generation (~2 um) in fibered Optical Parametric Chirped-Pulse Amplification. Indeed, in order to improve the spectral bandwidth of the PCF, we propose a microstructured air/silica fibre with a longitudinal gradient of its transverse section: a tapered fibre. During the fibre drawing, a homothety of the photonic structure allows to continuously vary the dispersion properties (including the zero-dispersion wavelength). This extends the usable range of the parametric gain over a spectral width of up to 100 nm, which is a key feature for generating few-cycle pulses in the MIR range.

The authors thank Agence National de la Recherche (ANR) for financing the METROPOLIS project (ANR-19-CE47-0008).

References

[1] H. Kano and H. Hamaguchi, "Characterization of a supercontinuum generated from a photonic crystal fiber and its application to coherent raman spectroscopy," Opt. Lett. 28, 2360 (2003).

[2] J.-C. Delagnes, R. Royon, J. Lhermite, et al., "High-power widely tunable ps source in the visible light based on four wave mixing in optimized photonic crystal fibers," Opt. Express 26, 11265 (2018).

[3] N. Valero, D. Marion, J. Lhermite, J.-C. Delagnes, W. Renard, R. Royon, and E. Cormier, "High-power amplified spontaneous emission pulses with tunable coherence for efficient non-linear processes," Scientific Reports (2021).

https://doi.org/10.1051/epjconf/202226702015

Optical Parametric Oscillator Based on Silicon Nitride Waveguides

Ming Gao[1], Niklas M. Lüpken[1], Klaus-J. Boller[2,1], Carsten Fallnich[1]

1. University of Münster, Institute of Applied Physics, Corrensstraße 2, 48149, Münster, Germany

2. University of Twente, MESA+ Institute for Nanotechnology, Enschede 7500AE, The Netherlands

Fiber-based optical parametric oscillators (OPOs) have been used for decades as wavelength-tunable light sources, which are attractive for applications such as coherent Raman spectroscopy or imaging [1,2]. However, the higher nonlinearity of integrated waveguides opens the potential to pave the way towards a fully-integrated waveguide-based OPO (WOPO) light source on a single chip [3,4]. Here, we study the WOPO exploiting four-wave mixing (FWM) in silicon nitride (Si_3N_4) waveguides synchronously pumped by an ultra-fast fiber laser.

The experimental setup consists of a Si_3N_4 waveguide (800 nm high, 1050 nm wide, and 10 mm long) pumped by a mode-locked fiber laser around an emission wavelength of 1030 nm with 40 MHz repetition rate to generate signal and idler sidebands by degenerate spontaneous FWM, shown in Fig.1 (a). Additionally, a Fourier filter behind the pump laser was used to adjust the pump pulse duration between 370 fs and 1 ps (not shown). The signal sideband was coupled into a polarization-maintaining fiber (PMF) and fed back to temporally and spatially overlap with the next pump pulse to stimulate the FWM process. Dispersive tuning of the WOPO was achieved by adjusting the free-space delay (Δt).

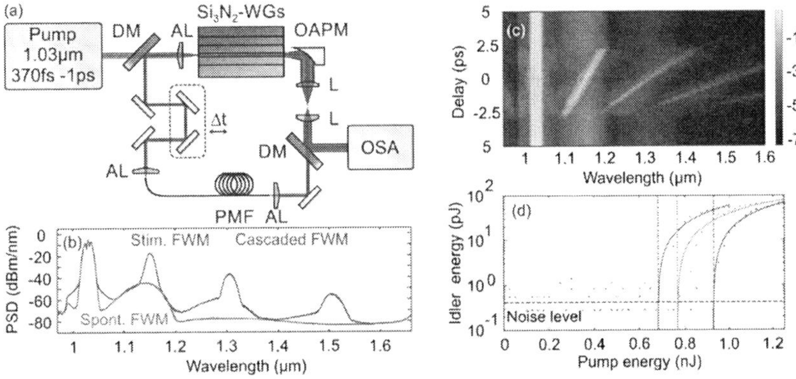

Fig.1 (a) Schematic of the experimental setup. (b) Spectra of spontaneous (red) and stimulated FWM (blue). (c) Color-coded output spectra of the WOPO as a function of the relative delay of the signal pulses at 1.25nJ pump energy. (d) External output energy of the WOPO as a function of external pump energy with different pump pulse durations (500, 700, and 900fs).

The presented WOPO showed 27 dB amplification in the idler sideband, followed by two stimulated cascaded FWM sidebands because of phase-matching between pump and idler waves. The tuning of the WOPO was accomplished by changing the free-space delay, thus, varying the temporal delay between stretched signal and pump pulses. As a result, the output idler wavelength was tuned from 1099 nm to 1194 nm, which corresponded to the bandwidth of the measured spontaneous FWM spectrum. The numerical calculations showed that, by engineering the waveguide dispersion, the accessible tuning range of the idler sideband could extend into 2 µm which is a compact pump source for various applications such as nonlinear microscopy. Moreover, to determine the oscillation threshold of the WOPO, the output idler energy versus pump energy was measured (Fig.1(d)), indicating an oscillation threshold at 0.68 nJ for 500 fs pump pulse duration, which increased with longer pump pulse duration. The maximum idler pulse energy of 63 pJ was extracted from the WOPO when pumped with 1.25 nJ at 900 fs.

In conclusion, compared to an earlier proof-of concept [4], this Si_3N_4-WOPO is even more advantageous for nonlinear microscopy and various other metrological applications by being operated at a 40 times higher repetition rate of 40 MHz, and is therefore representing an important step towards an all-integrated light source.

References

[1] M. Brinkmann, A. Fast, T. Hellwig, I. Pence, C. L. Evans, and C. Fallnich, "Portable all-fiber dual-output widely tunable light source for coherent Raman imaging," Biomed. Opt. Express **10**, 4437 (2019).

[2] K. Yang, S. Zheng, Y. Wu, P. Ye, K. Huang, Q. Hao, and H. Zeng, "Low-repetition-rate all-fiber integrated optical parametric oscillator for coherent anti-Stokes Raman spectroscopy," Opt. Express **26**, 17519 (2018).

[3] K.-Y. Wang, M. A. Foster, and A. C. Foster, "Wavelength-agile near-IR optical parametric oscillator using a deposited silicon waveguide," Opt. Express **23**, 15431 (2015).

[4] N. M. Lüpken, D. Becker, T. Würthwein, K.-J. Boller, and C. Fallnich, "Toward integrated synchronously pumped optical parametric oscillators in silicon nitride," Optics express **29**, 39895 (2021).

937 W Thulium:silica fiber MOPA operating at 2036 nm

Clément Romano[1], Dieter Panitzek[1], Dominik Lorenz[1,2], Patrick Forster[1,2], Marc Eichhorn[1,2], and Christelle Kieleck[1]

[1] *Fraunhofer IOSB (Institute of Optronics, System Technologies and Image Exploitation), Gutleuthausstraße 1, 76275 Ettlingen, Germany*
[2] *Institute of Control Systems (IRS), Karlsruhe Institute of Technology, Fritz-Haber-Weg 1, 76131 Karlsruhe, Germany*

Corresponding author: clement.romano@iosb.fraunhofer.de

Recent advances in high-power thulium-doped fiber lasers make them an attractive solution for applications in the 2 µm region, such as LIDAR, remote sensing, or defense applications. All these applications take advantage of a high transmission window in the atmosphere [1]. Output power greater than a kW have been demonstrated twice [2,3]. However, the rarity of similar results shows how challenging the power scaling of thulium-doped fiber lasers with output power greater than 600 W is, with primary constraints being heat management or thermal mode instability (TMI).

In this work, we present our latest results in power scaling in the 2 µm region. In order to maintain a stable atmospheric transmission a MOPA configuration is chosen instead of a single-oscillator configuration as it allows maintaining the wavelength over a wide range of output powers [4]. The MOPA configuration relies on a simple setup, consisting of a seed laser followed by a single-stage high-power amplifier. No isolator is used between the seed and the amplifier stage and operation relies only on the saturation of the amplifier by the seed laser. The seed laser is a monolithic thulium-doped fiber laser cavity based on fiber Bragg gratings. It delivers 10 W at 2036 nm with a linewidth of 0.1 nm at full-width half-maximum. The amplifier relies on a 25/400 thulium-doped fiber co-pumped by 793 nm multimode diodes. The setup is all-fiber spliced to allow for an efficient and robust design.

The graph on the left side of Figure 1 shows the output power versus pump power. The output power evolves linearly with the pump power and leads to a slope efficiency of 57.7 %. A maximum output power of 937 W was reached for a pump power of 1636 W. For the maximum output power this leads to an electric-to-optical efficiency of 24 %. The beam quality was measured over the whole pump power range and is plotted on the right hand side of Figure 1. The M^2 is close to diffraction limited, with values below 1.3. No TMI or other non-linear effects were observed up to the maximum output power.

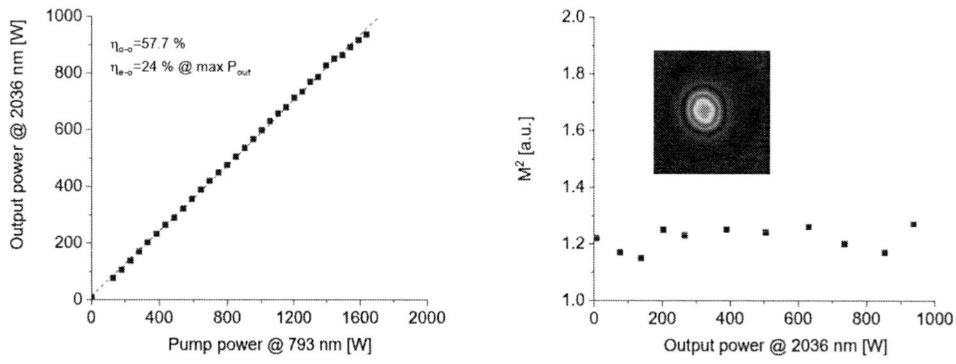

Figure 1: MOPA laser output power versus 793 nm pump power at an emission wavelength of 2036 nm.

This performance thereby demonstrates the readiness of the thulium-doped fiber laser technology for high-power applications. Operation of the MOPA was pump power limited; thus, further power scaling would be possible with brighter pump diodes as the TMI threshold is expected to be above 1 kW.

References

[1] McComb, T. S., et al. "Atmospheric Transmission Testing Using a Portable, Tunable, High Power Thulium Fiber Laser System," Proc. CLEO Paper JThJ5 (2010).
[2] Ehrenreich, T., et al. "1-kW , All-Glass Tm:fiber Laser," Proc. SPIE 7580 (2010).
[3] Anderson, B. M., et al. "Mode instability in kW-class thulium doped fiber amplifiers," Proc. SPIE 11981, 119810Y (2022).
[4] Walbaum, T., et al. "Monolithic thulium fiber laser with 567 W output power at 1970 nm," Opt. Lett. 41(11), p. 2632 (2016).

Efficient and Broadband Generation of Mid-Infrared Pulses by Optical Parametric Amplification in Dispersion-Engineered Thin Film Lithium Niobate

M. Hamrouni[1], A. Hwang[2], M. Jankowski[2,3], J. Mishra[2], H. S. Stokowski[2], T. P. McKenna[2], N. Jornod[2], C. Langrock[2], T. Südmeyer[1], A. H. Safavi-Naeini[2], M. M. Fejer[2]

1. *Laboratoire Temps-Fréquence, Université de Neuchâtel, Avenue de Bellevaux 51, Neuchâtel, Switzerland*
2. *E. L. Ginzton Laboratory, Stanford University, Stanford, California 94305, USA*
3. *NTT Research, Inc. Physics & Informatics Laboratories, 940 Stewart Drive, Sunnyvale, California 94085, USA*

Introduction. Optical parametric amplification (OPA) is an enabling technology for molecular spectroscopy and quantum photonics. Pumped by mature near-infrared lasers, OPA sources can generate broadband light in the mid-infrared (MIR) [1]. In many cases, low power requirements and compact footprints are crucial. However, in order to achieve sufficient parametric gain in standard nonlinear crystals, MIR OPA sources require driving lasers with high pulse energies and peak powers. These are typically provided by bulky and expensive ultrafast laser amplifiers. Here, we demonstrate a simple and efficient broadband MIR source based on OPA in thin-film lithium niobate (TFLN) on sapphire waveguides. These devices are directly driven by a commercially available mode-locked Yb-fiber laser and seeded by a CW telecom source to generate broadband pulses of MIR light. Confining light into nanophotonic TFLN waveguides allows for strong normalized efficiencies and the ability to engineer the waveguide dispersion by tuning its cross-sectional geometry. Here we dispersion-engineer symmetric temporal walk-off between the telecom signal and MIR idler near 3200 nm, with respect to the 1045-nm pump. Under these conditions, the large parametric gain centered around the peak of the pump pulse traps the signal and idler, thereby enabling OPA over distances much longer than the pulse-splitting length [2]. As a result, when pumped with picojoules of pulse energy, these devices efficiently generate short pulses of signal and idler. We observe ~60% of conversion efficiency for on-chip pump pulse energies of 7 pJ and measure up to 140 nm of idler bandwidth, which supports pulses as short as 75 fs. These devices represent one of the most compact sources of MIR light reported to date.

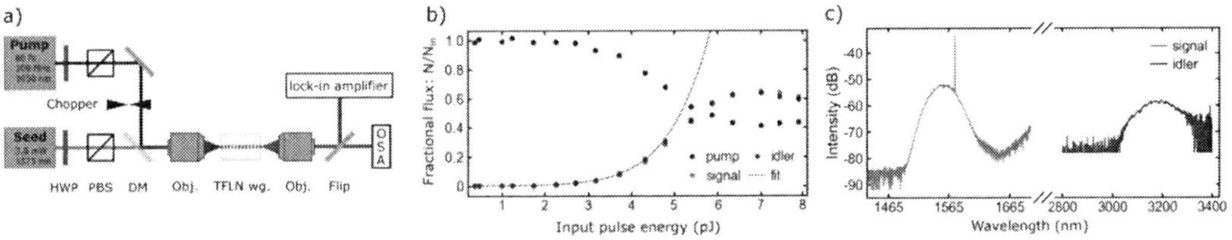

Fig. 1 a) Experimental setup. HWP, half-wave plate; PBS, polarizing beam-splitter; DM, dichroic mirror; Obj., reflective objective; TFLN wg., thin film lithium niobate waveguide; Flip, flip mirror; OSA, fiber-coupled optical spectrum analyzer. b) Generated signal and idler conversion efficiency (green and red dots, respectively). Dashed red: fit of the optical parametric growth. Fractional pump photon flux in blue. c) Generated signal and idler power spectral density in log scale for an input pulse energy of ~4 pJ. Both curves are offset for a better readability.

Experimental results. We use a 5-mm-long periodically poled TFLN nanowaveguide with a top width of 2885 nm, a film thickness of 1035 nm and an etch depth of 650 nm leading to a pump-signal and pump-idler group-velocity mismatch of +/- 20 fs/mm, respectively. The characterization setup is shown Fig. 1(a). The waveguide is pumped using a mode-locked laser delivering 80-fs pulses with a repetition rate of 100 MHz at a central wavelength of 1045 nm and seeded with 10 μW of on-chip average power from a CW laser emitting at 1575 nm. We chop the pump beam before the dichroic mirror [Fig. 1(a)] and use lock-in detection to obtain CW-background-free measurements of the average signal and idler output power generated by OPA. Under ~4.5 pJ of on-chip input pulse energy, the measured signal and idler output power are in a good agreement with the predicted exponential growth (Fig. 1(b), dashed red curve), reaching unsaturated gains as large as 37 dB. Above 4.5 pJ of pulse energy, the gain enters a saturated regime. In this regime, the depleted pump no longer traps the generated signal and idler and the three pulses walk off from each other [2]. As a result, using a maximum of 7.9 pJ of input pulse energy, the on-chip signal and idler average power reach up to 320 μW and 160 μW, respectively, corresponding to a conversion efficiency of 60% with respect to the on-chip pump photons. The measured signal and idler spectra [Fig. 1(c)] display a full-width at half maximum of approximately 37 nm and 140 nm, respectively, corresponding to sub-80-fs pulse durations assuming transform limited pulses.

Conclusion. We have demonstrated an efficient and broadband MIR source using OPA in dispersion-engineered TFLN nanophotonic waveguides. Using a commercially available pump and seed, we generated up to 160 μW of MIR average power and measured a 140-nm-broad idler spectrum centered at ~3200 nm. We believe this technology will open a new path for compact, efficient and broadband OPA.

References

1. G. Cerullo and S. De Silvestri, "Ultrafast optical parametric amplifiers," Rev. Sci. Instrum. **74**(1), 1–18 (2003).
2. C. Manzoni and G. Cerullo, "Design criteria for ultrafast optical parametric amplifiers," J. Opt. **18**(10), 103501 (2016).

Effect of multilayer substrate interference in planar waveguide scattering loss

Zhen Liu[1], Mohamed A. Ettabib[1], James S. Wilkinson[1], Michalis N. Zervas[1*]
1. Zepler Institute for Photonics and Nanoelectronics, University of Southampton, Southampton SO17 1BJ, UK

Introduction

Scattering loss due to a nanoscale interface roughness has been an unavoidable effect limiting planar waveguide performance, especially when the evanescent field at the core-cladding interfaces is strong, such as in waveguide-based sensors. Extensive theoretical work has been carried out to quantitively understand the scattering loss, among which Payne's approach [1] has been widely accepted. Following this work, Schmid et al [2] proved that the scattering loss is changed by interference between the two interfaces of a planar waveguide to a great extent, which is omitted in Payne's model. Since the loss can be significantly reduced due to interference, we have developed a theoretical model for scattering loss that takes account of the multilayer interference effect by extending Schmid's model. We have also fabricated a number of different Ta_2O_5 planar waveguides of different thicknesses deposited on Si wafers and compared the measured propagation loss to theory, showing a very good agreement.

Model and experiment

The analytical expression for scattering loss from TE modes is given as [2]

$$\alpha_{TE} = \frac{k_0^3}{4\pi n_{\text{eff}}}\left(n_2^2 - n_1^2\right)^2 \left(\mathrm{E}_0^x\right)^2 \int_0^\pi \left[\left(1 + r_s(\theta)\right)^2 + t_s^2(\theta)\right] \tilde{R}\left(\beta - n_2 k_0 \cos\theta\right) \mathrm{d}\theta$$

where E_0^x is the normalized TE mode modal field at the interface, n_{eff} is the effective refractive index of the waveguide mode, r_s and t_s are the total reflectivity and transmittance of the complete three-layer structure, $n_{2/1}$ is the refractive index of cladding/core. $\tilde{R}$ is the correlation function of the roughness and is proportional to the mean square of the surface fluctuation, σ^2. The expression for TM modes has a similar but more complicated form than TE and is also given by Schmid [2]. In our work, the transfer matrix method [3] is incorporated to calculate the r_s and t_s for waveguides with multilayer substrate.

To experimentally validate the model, Ta_2O_5 thin films with different thicknesses are deposited on silicon wafers as the waveguide core. The silicon wafer has a $2\mu m$ oxidized layer, so the planar waveguides are four-layer structures of cladding - Ta_2O_5 core- SiO_2- Si. Compared to the standard three-layer waveguide, the extra SiO_2/Si interface provides extra reflection, which has to be taken into account. The theoretical and experimental results are shown in Fig. 1. The theoretical values are calculated for a mean-square-roughness of one nanometer and the experimental values are normalized to the measured mean square of the surface fluctuation. In Fig. 1(a), by comparing Payne's and Schmid's model in the three-layer waveguide it is shown that the interference effects reduce the scattering loss significantly. In Fig. 1(b), the loss coefficients for the four-layer structure are calculated with our extended model and agree well with the experimental results. It is shown that the additional SiO_2/Si interface, while not affecting the field distribution of the guided modes, has a positive effect on the scattering loss. The experimental results have proven the accuracy of our extended model and imply that scattering loss could be reduced even further via interference effects in a purpose-designed multilayer substrate.

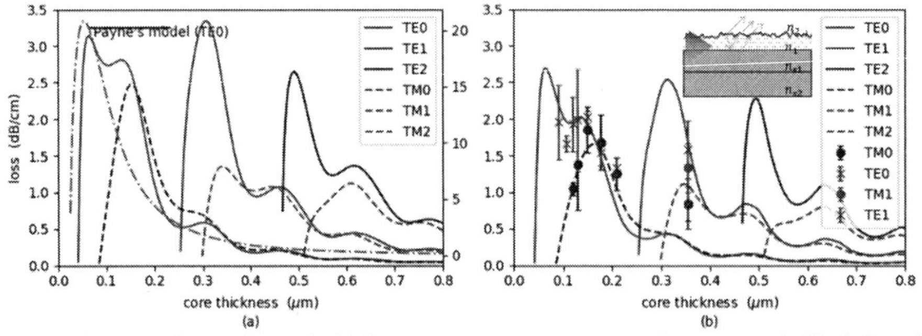

Fig. 1 Theoretical scattering loss coefficient for (a) three-layer air - Ta_2O_5 core - SiO_2, planar waveguide (the dash-dot line is scattering loss from Payne's model that omits the interference effect – plotted on the right Y-axis). (b) four-layer structure and the experimental results (the silica substrate layer is 2μm thick). The operation wavelength is 633nm.

References

[1] F. P. Payne and J. P. R. Lacey, A Theoretical Analysis of Scattering Loss from Planar Optical Waveguides, Optical and Quantum Electronics (1994), 26.
[2] J. H. Schmid, A. Delâge, B. Lamontagne, J. Lapointe, S. Janz, P. Cheben, A. Densmore, P. Waldron, D.-X. Xu, and K. P. Yap, "Interference effect in scattering loss of high-index-contrast planar waveguides caused by boundary reflections," Optics Letters 33(13), 1479–1481 (2008).
[3] S. J. Byrnes, "Multilayer optical calculations," arXiv:1603.02720 1–20 (2016).

Generation of 12 nJ Pulse Energy by a Thulium-doped Fiber Mamyshev Oscillator

***B. Schuhbauer*[1], *V. Adolfs*[1], *F. Haxsen*[1], *A. Wienke*[1,2], *U. Morgner*[1,2,3], *J. Neumann*[1,2], *D. Kracht*[1,2]**

1.Laser Zentrum Hannover e.V., Laser Development Department, Ultrafast Photonics Group, Hollerithallee 8, 30419 Hannover, Germany

2.Cluster of Excellence PhoenixD (Photonics, Optics, and Engeneering Innovation Across Disciplines), Hannover, Germany

3. Institut für Quantenoptik, Leibniz Universität Hannover, Welfengarten 1, 30167 Hannover, Germany

As a dynamic field in ultrafast fiber oscillators, Mamyshev oscillators proved pulse energy scaling capabilities, in particular in ytterbium- and erbium-based systems. Up to 190 nJ of pulse energy and compressed pulse durations below 50 fs have been achieved by using standard single mode step-index fibers in the spectral range of 1 µm [1,2].

Transferring this unique oscillator concept towards the near mid-infrared wavelength range was achieved in our group owing to the integration of passive normal dispersive fiber for dispersion-management and nonlinear spectral broadening, alongside with anomalous dispersive core-pumped gain fibers. This approach generated up to 6.4 nJ pulse energy with a compressed pulse duration of 156 fs [3]. However, these parameters remained only in the range of the best previously reported ultrafast thulium-doped fiber oscillators. The main limitation of the Mamyshev oscillator was excessive amplified spontaneous emission (ASE) in the short wavelength band of thulium around 1850 nm, which led to parasitic lasing and in turn to a destabilization of the mode-locked pulse train by gain competition.

Here, we address this limitation by a tailored red-shifted gain spectrum in one of the oscillator arms: Replacing one of the core-pumped gain fibers with a highly thulium-doped double-clad fiber resulted in a signal-to-ASE ratio of 22 dB owing to the re-absorption inside the fiber, which was an improvement by over an order of magnitude. The significant reduction of ASE allowed to generate up to 128 mW average power at a repetition rate of 10.25 MHz. Corresponding to the average power, 12 nJ of pulse energy were calculated at the output. In agreement to these high-energy pulses, the optical spectrum was governed by characteristic modulations owing to self-phase modulation and spanned about 60 nm, as presented in Figure 1a. The emitted up-chirped pulses could still be compressed to 154 fs, which is only 16 % above the transform limited pulse duration, as shown in Figure 1b.

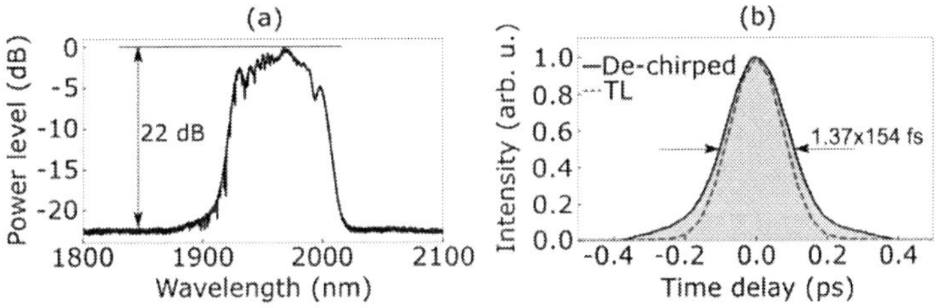

Fig. 1 Pulse characteristic in the spectral and temporal domain: a) Optical output spectrum and b) autocorrelation after compression and the corresponding transform limited pulse (TL).

To best of our knowledge, this is the highest reported pulse energy of an ultrafast thulium-doped fiber oscillator to date. Future improvements will include the replacement of the first arm by a passive normal dispersive arm, which makes the setup more compact and convenient to operate.

Acknowledgement
The authors gratefully acknowledge the support by the Deutsche Forschungsgemeinschaft (DFG, German Research Foundation) under Germany's Excellence Strategy within the Cluster of Excellence PhoenixD (EXC 2122, Project ID 390833453) and the ZIM program (grant No. 16KN053068).

References

[1] Zhanwei Liu, Zachary M. Ziegler, Logan G. Wright, and Frank W. Wise, "Megawatt peak power from a Mamyshev oscillator," Optica 4, 649-654 (2017)
[2] Pavel Sidorenko, Walter Fu, Logan G. Wright, Michel Olivier, and Frank W. Wise, "Self-seeded, multi-megawatt, Mamyshev oscillator," Opt. Lett. 43, 2672-2675 (2018)
[3] Paul Repgen, Benedikt Schuhbauer, Moritz Hinkelmann, Dieter Wandt, Andreas Wienke, Uwe Morgner, Jörg Neumann, and Dietmar Kracht, "Dispersion-managed thulium-doped fiber mamyshev oscillator," in Fiber Lasers XVIII: Technology and Systems, paper number 11665, Michalis N. Zervas, ed. (SPIE, 2021)

Numerical Analysis of Tapered Multicore Fibres for Laser System Scaling

Christopher Aleshire[1], Albrecht Steinkopff[1], Arno Klenke[1,2], Cesar Jauregui[1], and Jens Limpert[1,2,3]

1. Institute of Applied Physics, Friederich-Schiller-University Jena, Albert-Einstein-Straße 15, 07745 Jena, Germany

2. Helmholtz-Institute Jena, Fröbelstieg 3, 07743 Jena, Germany

3. Fraunhofer Institute for Applied Optics and Precision Engineering, Albert-Einstein Straße 7, 07745 Jena, Germany

Active multicore fibres (MCFs) enable compact and efficient scaling of laser power and energy, for example in coherent beam combination (CBC) schemes [1]. High-energy amplification was recently demonstrated with a tapered MCF which improved beam quality in large multimode cores, enhancing the scalability of these systems [2]. Tapering MCFs to achieve diffraction-limited beam quality with independent operation of the cores presents unique design challenges in addition to typical constraints of tapered fibres. In this submission, numerical analyses using Beam Propagation Method (BPM) are presented to guide the fabrication of tapered MCFs.

A microscope image of an in-house fabricated Yb-doped 4x4 MCF is shown in Fig. 1a. The cores in these fibres are densely arranged with a core separation of 2.5x the core diameter for high spatial overlap with the surrounding pump cladding and high pump absorption per unit length. The low core NA (0.04) enables nearly-single-mode operation at ~20-micron core diameter, with multimode operation in larger cores. However, the single-mode operating regime can be extended to much larger cores by tapering one end of the MCF to allow only fundamental-mode excitation. In general, tapered fibre amplifiers transition gradually from a single-mode input to a multimode output over several metres [3]. In MCFs, long taper lengths can result in significant inter-core cross-talk, particularly in the single-mode region where the evanescent overlap of the core fields is greatest. However, the minimum taper length is limited by mode-mixing [4] and bend loss due to the non-axiosymmetric nature of MCF tapers. BPM simulations of MCFs tapered from 15- to 25-micron core diameter in Fig. 1 b-c show these effects for a 5 mm taper length (Fig. 1c) which exhibits mode mixing and bend loss, and for a 50 mm taper length (Fig. 1d) which exhibits inter-core cross-talk. For this fibre configuration, a working taper design space can be defined which minimizes mode distortion and/or crosstalk depending on the target application.

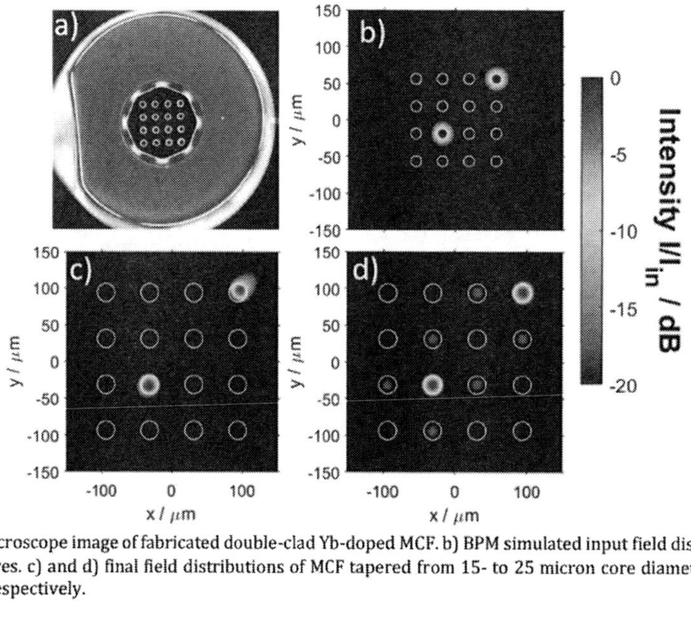

Fig. 1 a) Microscope image of fabricated double-clad Yb-doped MCF. b) BPM simulated input field distribution with 15 micron cores. c) and d) final field distributions of MCF tapered from 15- to 25 micron core diameter over 5 mm and 50 mm, respectively.

Numerical techniques such as BPM can accurately model bend loss, intra-core mode-coupling, and inter-core cross-talk in complex tapered MCF structures. The results of these simulations can guide the tapering of existing high-power compatible MCF designs as a post-processing step as shown here, and further applied for the design of "taper-ready" MCFs. Thus the mode-area scaling property of tapered gain fibres can be utilized in active MCFs to further optimize a new generation of high-power coherently combined laser systems.

[1] A. Klenke et al. "500 W rod-type 4 × 4 multicore ultrafast fiber laser," Opt. Lett. 47, 345-348 (2022).
[2] C. Aleshire et al. "High-energy Q-switched 16-core tapered rod-type fiber laser system," Opt. Lett. 47, 1725-1728 (2022).
[3] J. Kerttula et al. "Tapered fiber amplifier with high gain and output power," Laser Phys. 22, 1734–1738 (2012).
[4] J.D. Love et al. "Tapered single-mode fibres and devices. Part 1: Adiabaticity criteria," IEE Proc.-J. 138, 343-354 (1991).

 https://doi.org/10.1051/epjconf/202226702021

Crystalline Grating-Waveguide Resonant reflectors

G. Mourkioti[1], G.A. Govindassamy[1], F. Li[2], R.W. Eason[1], M. Abdou Ahmed[3], J.I. Mackenzie[1]
[1]Optoelectronics Research Centre, University of Southampton, Southampton, SO17 1BJ, UK; [2]University of Eastern Finland, FI-80100 Joensuu, Finland; [3] Institut für Strahlwerkzeuge, University of Stuttgart, 70569 Stuttgart, Germany

The demand for reliable polarised narrow-bandwidth high-power lasers has motivated the development of innovative diffractive optics that can meet such requirements. Grating-Waveguide Structures (GWS) combining a waveguide and sub-wavelength grating can be used as a polarisation-selective high-efficiency reflector [1]. However, the maximum power (fluence) that can be used with these optics is limited by the intrinsic absorption and damage threshold of the compound materials. In this work, we report on the first resonant reflectors based on GWS made entirely of crystalline materials, which potentially offer better thermal and mechanical properties. Such crystalline GWS devices could enable operation at higher powers than conventional materials currently allow.

Two GWS designs were investigated in this study, targeting operation at 1030 nm and 1970 nm, to facilitate polarised wavelength tuning of Yb- or Tm-lasers. These devices comprise a sub-wavelength grating etched into a sapphire substrate on which an epitaxially grown waveguide layer of Sc_2O_3 is deposited (Fig.1a). In our experiments, 4-inch (0001) sapphire wafers were used as the substrate. For the respective designs, a 200-nm or 400-nm SiO_2 layer was deposited to act as hard mask to dry etch the substrate. An extra 50-nm Cr layer was deposited on top of SiO_2, which acts as conductive layer in the patterning process, but also as hard mask for the dry etching of SiO_2. E-beam lithography was used to pattern linear gratings with a 515-nm or 984-nm period at 50% duty cycle, for the 1-and 2-μm devices respectively. Inductively Coupled-Plasma (ICP) etching with Cl_2/O_2 gas chemistry was used to etch the Cr layer, followed by O_2 plasma cleaning to remove any residual resist. ICP with CHF_3/Ar was then used to etch the SiO_2 layer followed by Cl_2-based dry etching to remove the residual Cr layer. At this stage the sapphire substrate is patterned with a 200-nm or 400-nm thick SiO_2 mask. Sapphire structuring was achieved using ICP with $BCl_3/Cl_2/Ar$ or $SiCl_4/Cl_2/Ar$ gas chemistry (Fig.1b top), followed by immersion in hydrofluoric acid (7:1) to remove the residual SiO_2. Dicing out a ~1-cm^2 chip from the wafer, the final step in the waveguide reflector fabrication process is the growth of Sc_2O_3 on top of the sapphire grating using pulsed-laser deposition (PLD). The structured chip was heated to ~1100° C by a CO_2 laser, in chamber with an O_2 background gas pressure of 20 μbar. A 100-Hz KrF excimer laser ablated a ceramic Sc_2O_3 target, positioned 55mm in front of the heated substrate. Optimisation of growth conditions was made on flat sapphire substrates to determine the growth rate and film properties [2].

SEM images of the GWS are shown in Fig.1b(top) with residual SiO_2 mask before its removal and Fig.1b(bottom) after PLD. To characterise the performance of the crystalline reflectors their transmission was measured with a 1030-nm linearly polarised laser beam in function of the incident angle (Fig. 1c). For both TE- and TM-polarisations, a drop in the sample's transmission to ~40% was observed at angles of incidence of 6.1° and 7.4° respectively, demonstrating coupling of the incident beam to the waveguide mode. With a corresponding measured reflectance of 30% (TE) and 47% (TM), whereby the deficit is due to losses, either in the material or scattering. A discrimination of 5:1 for the TM-resonance was found, which could be improved with a back-surface AR coating.

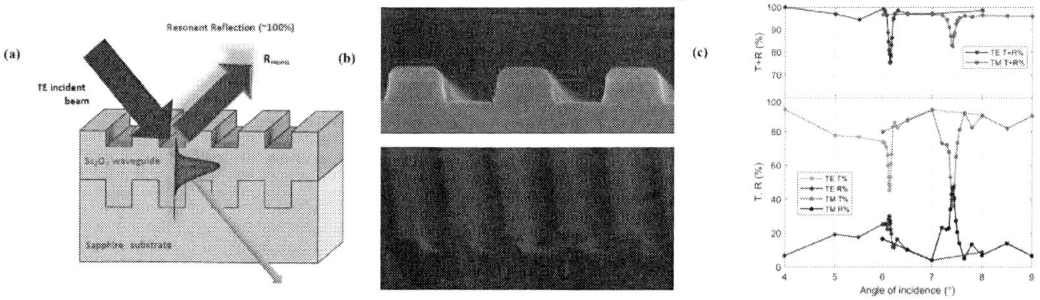

Fig. 1 (a) Grating waveguide reflector concept; **(b)** SEM images of structured SiO_2-masked sapphire (top) and after Sc_2O_3 deposition (bottom); and **(c)**TE/TM-polarised resonance characterisation at 1030 nm, dips in sum of transmittance and reflectance equates to losses.

In summary, we have fabricated and characterised the first crystalline reflectors based on GWS, with a resonant reflectivity at 1-micron of 47% achieved. Improvement in the surface quality of the crystalline GWS is required to achieve the ideal ~100% resonance reflection efficiency, as material/scattering losses appear to limit current performance. Refinement of the fabrication processes will lead to better device efficiency for these narrow-linewidth polarisation-selective reflectors.

References

[1] D. Rosenblatt, A. Sharon, A.A. Friesem, "Resonant grating waveguide structures", IEEE J. Quantum Elect., **33** (11), 2038-2059 (1997).
[2] G.A. Govindassamy, J.J. Prentice, J.G. Lunney, R.W. Eason, J.I. Mackenzie, "Effect of laser repetition rate in the growth of Sc_2O_3 via pulsed laser deposition", accepted in Appl. Phys. A-Mater. in May 2022.

Self-phase modulation in periodically-poled thin-film lithium niobate waveguides

Gamze Gul[1], Kamal Abdelsalam[2], Sasan Fathpour[2,3], Kim F. Lee[4], Gregory S. Kanter[4], and Prem Kumar[1,4]

1. *Graduate Program in Applied Physics, Northwestern University, 2145 Sheridan Road, Evanston, Illinois 60208, USA*
2. *CREOL, The College of Optics and Photonics, University of Central Florida, Orlando, Florida 32816, USA*
3. *Department of Electrical and Computer Engineering, University of Central Florida, Orlando, Florida 32816, USA*
4. *Center for Photonic Communication and Computing, Department of Electrical and Computer Engineering, Northwestern University, Evanston, Illinois 60208, USA*

Thin-film lithium niobate (TFLN) waveguides have been one of the most popular elements for integrated photonics in recent years. The design of the waveguide and the high refractive index contrast create strong field confinement and enable low power nonlinear interactions. The periodic-poling provides quasi-phase-matching (QPM) for the second-order ($\chi^{(2)}$) nonlinear processes. Periodically-poled TFLN waveguides have been employed in various applications such as supercontinuum generation [1], third-harmonic generation [2], and photon-pair generation [3] in a much more efficient way with low powers due to the cascaded $\chi^{(2)}$ effect. Although cascaded interactions are more effective near the phase-matching wavelength, their effect can still be observed through self-phase modulation, which is originally a third-order ($\chi^{(3)}$) nonlinear effect [4,5].

Here we investigate the self-phase modulation (SPM) of sub-picosecond telecom wavelength light pulses due to the cascaded second-order nonlinear processes in periodically-poled TFLN waveguides when the spacing between the wavelength of the input pulse and the phase-matching wavelength is large. The experiments are performed on x-cut 1-cm long waveguides with a 1-mm long poled region in the middle with grating periods of 2.7 and 2.76 μm, respectively. The free-space output from an IMRA Femtolite laser is coupled into the waveguides with a 40x microscope objective to avoid any dispersion or nonlinear effects on the pulses before the waveguide. We collect the output light with a lensed fiber and calculate the change in the spectral bandwidth $\Delta\omega_{\mathrm{rms}}$. We also used a split-step Fourier method to simulate the propagation of pulses in the presence of nonlinearities in the waveguides, the lensed fiber, and fiber patch cords in the setup under the condition of an undepleted pump.

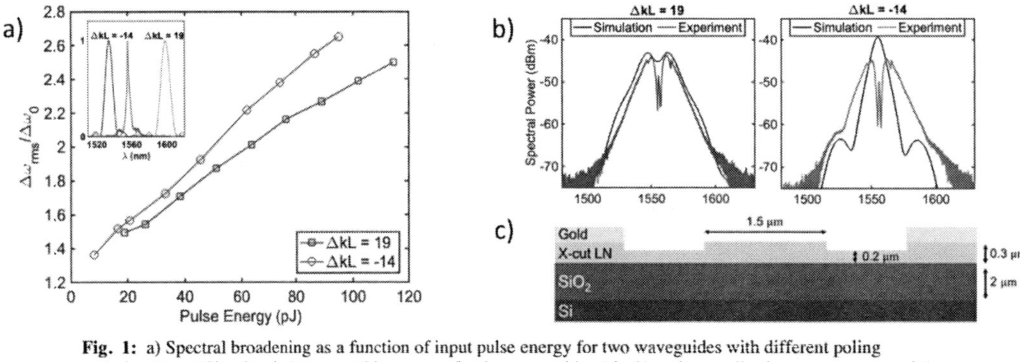

Fig. 1: a) Spectral broadening as a function of input pulse energy for two waveguides with different poling periods. (Inset) Simulated phase matching curves for the waveguides (black) and normalized power spectrum of the input laser (red). b) Simulated and experimentally observed output spectra of pulses for 40 pJ input pulse energies. c) Cross-section of the ridge TFLN waveguide.

We experimentally studied the spectral broadening in periodically-poled TFLN waveguides at telecom wavelengths. We observed a 2.6-factor spectral broadening for <90 pJ when phase-mismatching is negative (cf. Fig. 1(a)). This value decreases to 2.3 for the other waveguide with positive phase-mismatching. Figure 1(b) shows that the simulation performed with a 400 fs FWHM sech^2 type pulse agrees with the experimentally observed spectra for the lower pulse energies whereas in the middle of the spectra it fails to match the experimental spectra. The effect of cascaded $\chi^{(2)}$ on spectral broadening can be further analyzed by adding $\chi^{(3)}$ effects to the simulation. Such details and more experimental results under various operating conditions will be presented at the conference.

References

[1] M. Jankowski *et al.*, "Quasi-static optical parametric amplification," Optica **9**, 273 (2022).
[2] A. S. Solntsev *et al.*, "Cascaded third harmonic generation in lithium niobate nanowaveguides," Appl. Phys. Lett. **98**, 231110 (2011).
[3] B. S. Elkus, K. Abdelsalam, S. Fathpour, P. Kumar, and G. S. Kanter, "Quantum-correlated photon-pair generation via cascaded nonlinearity in an ultra-compact lithium-niobate nano-waveguide," Opt. Express **28**, 39963 (2020).
[4] R. H. Stolen and C. Lin, "Self-phase-modulation in silica optical fibers," Phys. Rev. A **17**, 1448 (1978).
[5] R. Schiek, "Self-phase modulation in lithium niobate waveguides," in 2014 International Conference Laser Optics (2014).

https://doi.org/10.1051/epjconf/202226702023

Optimization of the temporal quality of ultrafast pulses using dispersion scan based on tunable chirped fiber Bragg gratings

Meng Liu[1,2], Ayhan Tajalli[1], Mingqi Fan[1], Christoph Mahnke[1], Jiaan Zheng[1], Ingmar Hartl[1]

1. Deutsches Elektronen-Synchrotron DESY, Notkestraße 85, 22607 Hamburg, Germay
2. South China Normal University, Guangzhou, Guangdong 510006, China

The XUV Free-Electron Laser (FEL) facility FLASH at DESY provides ultrashort pulses of high-intensity radiation in the extreme ultraviolet and soft X-ray range, which is used as a tool for ultrafast dynamics in molecular physics, chemistry, biology, material science. Vast majority of applications, seeding, and driving electrons from RF photocathode guns of FELs require high power ultrashort laser pulses. Recently, within the upgrade program FLASH2020+, an improved Yb:fiber picosecond front end was developed for seeding high power Yb:YAG gain blocks. Using chirped pulse amplification technology, the pulses are stretched first to 650 ps duration and compressed after amplification to achieve high power transform-limited sub-picosecond pulses for OPCPA pumping or multi-pass cell compression.

In our upgraded system design, we plan for improved stability in 24/7 operation to fix the grating compressor components after initial alignment and optimize the pulse quality as well as correct for any residual day-to-day dispersion drifts (caused by for example environmental factors) via tunable chirped fiber Bragg gratings (CFBG) used for stretching the oscillator pulses. As we know, the dispersion scan (d-scan) technique [1] can be employed to characterization and optimization of the temporal pulse quality. The transform-limited pulse duration in our system is ~550 fs, which means that glass wedges which are typically used in the d-scan technique can't introduce sufficient dispersion for suitable d-scan traces. One option would be to scan the grating compressor dispersion [2,3], which is however not compatible with our goal of a fixed compressor set-up.

Here we present a different option which does not require to use any movable parts for d-scan but utilizes the tunability of our CFBGs (Teraxion) via distributed heaters. The laser schematic is shown in Fig. 1(a). The output pulses from the oscillator are stretched by a tunable CFBG 1. After a pre-amplifier, the repetition rate is down counted to 1 MHz. Then, the pulses are split into two branches. The main branch will be further stretched by tunable CFBG 2, amplified and down-counted to 100 kHz. Note that in our demonstration set-up the last amplifier is a Yb:YAG bulk amplifier, where the output pulse duration is ~650 ps and average power is ~2.5 W. In the final set-up we will further amplify this output to 400 W power. By coarse alignment of the compressor, a pulse duration of ~750 fs could be achieved. D-scan traces are generated by tuning the CFBGs and recording second harmonic spectra using a thin BBO crystal. Figures 1(b) and 1(c) depict two d-scan traces, where the 2nd order dispersion is scanned with fixed but different amounts of 3rd order dispersion. The corresponding reconstructed autocorrelation traces at the GDD value corresponding to the shortest pulse are shown with red and blue curves in Figs. 1(d) and 1(e), respectively. The black curves show the measured autocorrelation traces at these points. The close to perfect consistency between measured and reconstructed autocorrelation traces prove the validity of our method. We also validated this novel d-scan technique by comparing the results to more traditional a grating compressor scan.

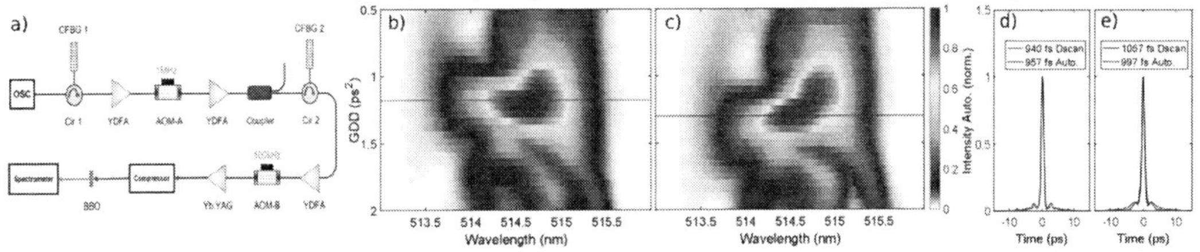

Fig. 1 Schematic layout of laser system together with d-scan and autocorrelation measurements of optimally and non-optimally compressed pulses after a grating compressor. (b) and (c) measured d-scan traces with different amount of introduced residual TODs by CFBG. (d) and (e) comparison of autocorrelation measurements (black curves) and reconstructed autocorrelations traces from d-scan measurements at optimum compression points (b) and (c), respectively. The values correspond to the width of the autocorrelation trace. The retrieved pulse durations from d-scan method are 665 fs and 747 fs for d) and e), respectively.

In conclusion, we demonstrated sub-picosecond pulse optimization using a d-scan technique based on tunable CFBGs. We used this technique to compress 650 ps pulses to ~650 fs with optimized pedestals. This method allows reliable pulse-characterization and optimization without movable parts and therefore no risk of mis-alignment of a laser system used in 24/7 operation at a large scale FEL facility.

References

[1] M. Miranda, T. Fordell, C. Arnold, A. L'Huillier, H. Crespo, "Simultaneous compression and characterization of ultrashort pulses using chirped mirrors and glass wedges," Opt. Express **20**, 688 (2012).
[2] B. Alonso, I.J. Sola, H. Crespo, "Self-calibrating d-scan: measuring ultrashort laser pulses on-target using an arbitrary pulse compressor," Sci. Rep. **8**, 3264 (2018).
[3] P. Repgen, A. Tajalli, D. Wandt, J. Neumann, D. Kracht, U. Morgner, "Complete characterization of ultrafast pulses of an Yb-doped fiber amplifier via dispersion scans after compression in a grism compressor," Proceedings **10512**, Fiber Lasers XV: Technology and Systems; 1051213 (2018).

Synchronously Pumped Tantalum Pentoxide Waveguide-based Optical Parametric Oscillator

Maximilian Timmerkamp[1], Niklas M. Lüpken[1], Shiqiprim Adrian Abazi[2], Julian Rasmus Bankwitz[2], Carsten Schuck[2], Carsten Fallnich[1]

1. *University of Münster, Institute of Applied Physics, Corrensstraße 2, 48149 Münster, Germany*
2. *University of Münster, Institute of Physics, Wilhelm-Klemm-Str. 10, 48149 Münster, Germany*

Optical parametric oscillators (OPOs) and especially fiber-based OPOs are versatile light sources for a variety of applications in, e.g., nonlinear microscopy [1]. To further miniaturize OPOs, integrated waveguides can be used as the nonlinear gain medium [2]. Tantalum pentoxide waveguides have so far attracted relatively low attention within nonlinear optics, although they offer a ten times higher nonlinearity than the well-established silicon nitride platform, while featuring comparable low loss and high mode-confinement [3]. Here, we present a synchronously pumped waveguide-based OPO (WOPO) employing tantalum pentoxide.

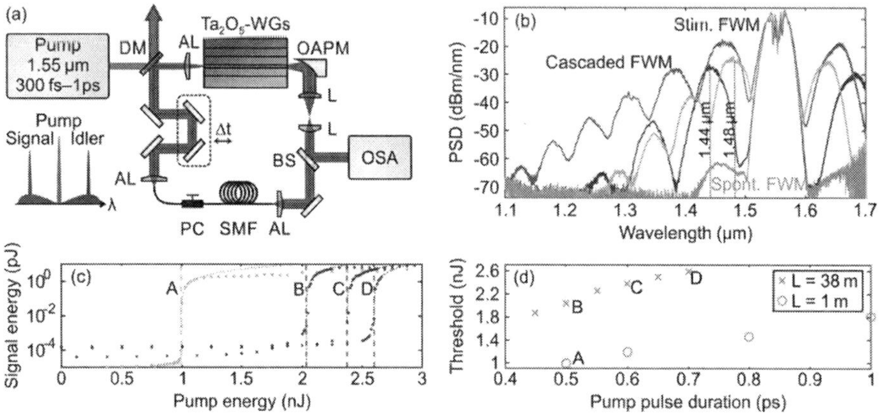

Fig. 1 (a) Schematic experimental setup. (b) Spectra of spontaneous (gray) and stimulated FWM (colored) for three different delays Δt (compare (a)). (c) Energy of the signal sideband (dots) as a function of the pump energy and linear fits (dotted) to determine the oscillation threshold (dashed) for four different parameter sets of pump pulse duration and feedback fiber length, as indicated in (d). (d) Oscillation threshold as a function of pump pulse duration.

We fabricated ridge waveguides and used a 0.6 μm high, 1.7 μm wide and 10 mm long waveguide as the nonlinear gain medium. Fig. 1(a) shows that the waveguide was pumped at 1.55 μm center wavelength with 80 MHz repetition rate. The pump pulse duration was adjusted by means of a Fourier filter. Inside the waveguide, signal and idler sidebands were generated via spontaneous four-wave mixing (FWM). The signal sideband was temporally stretched in a fiber, overlapped with the next pump pulse, and fed back into the waveguide to stimulate the FWM process, as can be seen in Fig. 1(b).

In order to investigate the performance of the WOPO, the length of the feedback fiber was initially chosen to be 1 m, corresponding to one signal pulse oscillating in the resonator. Fig. 1(c) shows an oscillation threshold of 1 nJ for a 500 fs long pump pulse. But the operation range in which the signal energy linearly increased with the pump energy was limited due to the system shifting into a regime of seeded supercontinuum generation (SCG) and, therefore, the maximum reachable signal power was 1 pJ.

To increase the output energy by circumventing SCG that was caused by the insufficient stretching of the signal pulse in the feedback fiber, the fiber length was increased to 38 m (15 pulses in the resonator). In this case, the oscillation threshold increased to 2 nJ at 500 fs pump pulse duration, however, also the signal output energy increased to 6 pJ due to avoiding the entering of the SCG regime by dispersive pulse stretching. Fig. 1(b) shows that the signal sideband's amplification was 40 dB over spontaneous FWM and the signal wavelength was freely tunable in the range between 1.44 μm and 1.48 μm.

In conclusion, we showed a synchronously pumped Ta_2O_5 waveguide-based optical parametric oscillator, that is to our knowledge the first of its kind. The Ta_2O_5-WOPO shows the potential to be fully integrated on a chip for applications in, e.g., optical metrology.

References

[1] M. Brinkmann, A. Fast, T. Hellwig, I. Pence, C. L. Evans, and C. Fallnich, "Portable all-fiber dual-output widely tunable light source for coherent Raman imaging," Biomed. Opt. Express **10,** 4437 (2019).

[2] N. M. Lüpken, D. Becker, T. Würthwein, K.-J. Boller, and C. Fallnich, "Toward integrated synchronously pumped optical parametric oscillators in silicon nitride," Opt. Express **29**, 39895 (2021).

[3] L. Splitthoff, M. A. Wolff, T. Grottke, and C. Schuck (2020). "Tantalum pentoxide nanophotonic circuits for integrated quantum technology," Opt. Express **28**, 11921 (2020).

100W, 1 mJ, few-cycle pulses at 2 μm wavelength

Ziyao Wang[1*], Tobias Heuermann[1,2,3], Martin Gebhardt[1,2,3], Mathias Lenski[1], Philipp Gierschke[1,4], Robert Klas[1,2,3], Cesar Jauregui[1] and Jens Limpert[1,2,3,4]

1. Institute of Applied Physics, Abbe Center of Photonics, Friedrich-Schiller-University Jena, Albert-Einstein-Str. 15, 07745 Jena, Germany

2. Helmholtz-Institute Jena, Fröbelstieg 3, 07743 Jena, Germany

3. GSI Helmholtzzentrum für Schwerionenforschung, Planckstraße 1, 64291 Darmstadt, Germany

4. Fraunhofer Institute for Applied Optics and Precision Engineering, Albert-Einstein-Str. 7, 07745 Jena, Germany

**Corresponding author: ziyao.wang@uni-jena.de*

In recent years, few-cycle laser sources with high pulse energy and high repetition rate in the short-wavelength infrared (SWIR) region (from 1.4 μm to 3 μm) have become increasingly impactful for scientific and industrial applications [1, 2]. Compared to the well-developed 1 μm technologies, the longer wavelength possesses fundamental advantages for nonlinear frequency conversion processes, such as mid-infrared [3], THz [4], and soft X-ray [5] generation. One common method to address the SWIR region is optical parametric chirped pulse amplification (OPCPA), which already demonstrated direct generation of sub 20 fs pulses [5] or more than 100 W average power [6] with mJ-level pulse energies. An alternative method is to post-compress the output of thulium(Tm)-doped fiber laser systems, which have been proven to be a viable approach for power scaling in the SWIR region, reporting 1 kW of average output power [7] or 15 GW of pulse peak power [8] in ultrafast operation. In this contribution, we present our first results of nonlinear post compression of a high-power Tm-doped fiber laser in a gas-filled capillary. The laser source is based on the coherent combination of four ultrafast Tm-doped fiber amplifiers, delivering a combined pulse energy of 1.67 mJ, with 85 fs pulse duration at a repetition rate of 101 kHz [8]. The post compression stage consists of a 63 cm long rod-type hollow-core fiber with an inner diameter of 350 μm, which is placed on a V-groove inside a high pressure gas vessel filled with ~2 bar of argon. The output pulses are recompressed using ~3 mm of fused silica. In this experiment, we launched an average power of 145 W to the hollow core fiber. Fig. 1 shows the compression results. The output spectrum is measured with a multi-mode fiber. The compressed pulses are analyzed by a commercial TIPTOE device (company: SourceLab) [9]. We achieved 102.8 W of average output power and a compressed pulse duration of 17.6 fs (2.75 cycles), corresponding to an output pulse energy of 1.01 mJ and an estimated pulse peak power of 35.8 GW.

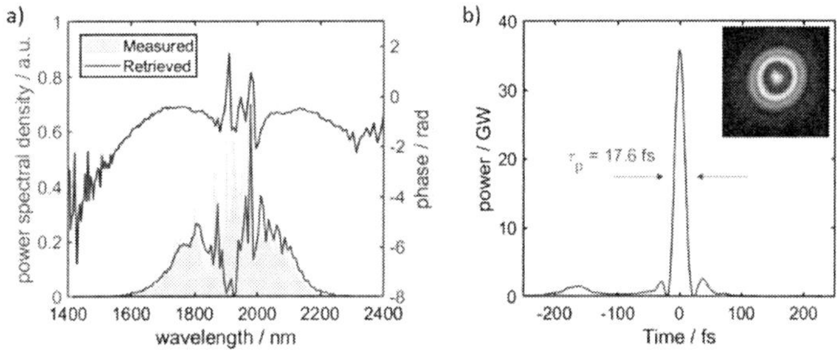

Fig. 1 Results of the nonlinear compression. a) measured spectrum (light blue), retrieved spectrum (dark blue) and retrieved phase (red). b) retrieved pulse profile. The inset in b) corresponds to the near-field image of the beam at the output of the hollow-core fiber.

To the best of our knowledge, this system represents the highest average output power mJ-class few-cycle source in the SWIR to date. Given its unique parameter set, this source can serve as the ideal platform for nonlinear frequency conversion into the soft X-ray, mid infrared and THz regions. Furthermore, we believe that the scalability of coherent beam combination of Tm-doped fiber lasers combined with hollow-core fiber compression will enable multi-mJ, multi 100 W, few-cycle output at 2 μm wavelength in the near future.

References

[1] Gebhardt et al., "Bright, high-repetition-rate water window soft X-ray source enabled by nonlinear pulse self-compression in an antiresonant hollow-core fibre," Light Sci Appl **10**, 36 (2021).
[2] Mingareev et al., "Principles and applications of trans-wafer processing using a 2-μm thulium fiber laser," Int J Adv Manuf Technol **84**, 2567–2578 (2016).
[3] Butler et al., "Watt-scale 50-MHz source of single-cycle waveform-stable pulses in the molecular fingerprint region," Opt. Lett. **44**, 1730-1733 (2019).
[4] Vicario et al., "High efficiency THz generation in DSTMS, DAST and OH1 pumped by Cr:forsterite laser," Opt. Express **23**, 4573-4580 (2015).
[5] Pupeikis et al., "Water window soft x-ray source enabled by a 25 W few-cycle 2.2 μm OPCPA at 100 kHz," Optica **7**, 168-171 (2020).
[6] Windeler et al., "100 W high-repetition-rate near-infrared optical parametric chirped pulse amplifier," Opt. Lett. **44**, 4287-4290 (2019).
[7] Gaida et al., "Ultrafast thulium fiber laser system emitting more than 1 kW of average power," Opt. Lett. **43**, 5853-5856 (2018)
[8] Heuermann et al., "Ultrafast Tm-doped fiber laser system delivering 1.65 mJ, sub-100 fs pulses at 100 kHz repetition rate," accepted for publication in Optics letters, DOI 10.1364/OL.459385
[9] Cho et al., "Reconstruction algorithm for tunneling ionization with a perturbation for the time-domain observation of an electric-field," Sci Rep **11**, 13014 (2021).

Fiber-tip nanothermometer based on up-conversion nanocrystals for electrolysis cells

Lea Kötters[1,5], Simon Spelthann[1], Lena Bühre[2], Rajesh Komban[3], Horst Weller[3,4], Richard Hanke-Rauschenbach[2], Detlev Ristau[1,5,6], Christoph Gimmler[3], Boris Bensmann[2], Michael Steinke[1,6]

1. Leibniz University Hannover, Institute of Quantum Optics, Welfengarten 1, D-30167, Hannover, Germany

2. Leibniz University Hannover, Institute of Electric Power Systems, Appelstraße 9A, D-30167 Hannover, Germany

3. Fraunhofer Center for Applied Nanotechnology CAN – (a research division of Fraunhofer Institute for Applied Polymer Research), Grindelallee 117, D-20146 Hamburg, Germany

4. University Hamburg, Department of Chemistry, Martin-Luther-King-Platz 6, D-20146 Hamburg, Germany

5. Cluster of Excellence PhoenixD (Photonics, Optics, and Engineering – Innovation Across Disciplines), Welfengarten 1A, D-30167, Hannover, Germany

6. Leibniz University Hannover, QUEST-Leibniz-Research School, Callinstraße 36, D-30167 Hannover, Germany

Characterisation of *Proton Exchange Membrane Water Electrolysis* (PEMWE) cells requires the knowledge of the state variables such as concentrations, temperatures, and pressure inside the cell. For example, information about the temperature at the catalyst layer will help to better parametrize reaction kinetics and understand degradation mechanisms. However, PEMWE cells are operated under strong compression forces to obtain the required contact pressure. The resulting closed setups make it difficult to determine the state variables inside the cell and often only averaged values obtained from the cells inlet and outlet are available, although strong gradients inside the cell can be expected. To overcome this issue by providing a suitable sensor, we developed a novel, miniature fibre-tip sensor for proof-of-concept nanothermometry inside PEMWE cells as outlined in the following.

The Boltzmann population of the Stark energy levels of rare earth ions in crystalline hosts provides a temperature dependent emission. This feature allows to measure temperature optically [1]. In particular, Er^{3+}:Yb^{3+} co-doped $NaYF_4$ UC-NCs can conveniently be excited around 980 nm with standard laser diodes and emit via up-conversion in the green wavelength regime. This green emission features multiple narrow lines whose relative intensities allow (after a proper calibration) to compute the NC temperature [2]. We used an optical approach [3] in combination with simple dip-coating to attach UC-NCs to the 230 µm fiber facet of the output of a 2-to-1 multimode fiber splitter, see Fig. 1 (a). The splitter separates the incoming excitation radiation provided by a standard 976 nm laser diode and the backward propagating green emission from the UC-NCs, which is read-out by a USB-spectrometer. The UC-NCs have a core/shell structure with an average core size of 14 nm and a shell thickness of around 1.5 nm, see Fig. 1 (b) for a TEM image. The relative Er^{3+}:Yb^{3+} doping concentration is around 1:10. The sensor has been calibrated by controlled heating of the fiber tip from 50 °C to 90 °C for various integration times of the data acquisition via the USB-spectrometer. At a constant sensitivity of 0.81 %/K, an integration time of 10 s or 150 s yielded a temperature resolution of 1.0 mK or 0.2 mK, respectively, see Fig. 1 (c). In preliminary experiments, we used the fiber-tip sensor for in-situ temperature measurements in PEMWE cells, see Fig. 1 (d). In our contribution, we will present the development and calibration of the sensor as well as its application in the PEMWE cell.

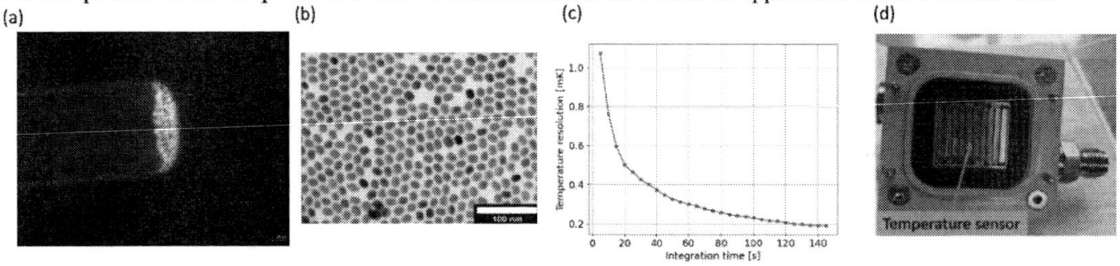

Fig. 1 (a): Microscopy image of the fiber-tip (diameter: 230 µm) with attached UC-NCs and their characteristic green emission. (b): TEM image of the used UC-NCs with an average core size of 14nm. (c): Absolute temperature resolution of the sensor in dependency of the integration time. (d): PEMWE with drilled entrance hole for the fiber-tip sensor.

Literature

[1] J. Thiem, S. Spelthann, L. Neumann, F. Jakobs, H.-H. Johannes, W. Kowalsky, D. Kracht, J. Neumann, A. Ruehl, D. Ristau, „Upconversion Nanocrystal Doped Polymer Fiber Thermometer", Sensors **20**, 6048 (2020)

[2] F. Vetrone, R. Naccache, A. Zamarron, A. Fuente, F. Sanz-Rodriguez, L.M. Maestro, E.M. Rodriguez, D. Jaque, J. G. Sole, J. A. Capobianco, „Temperature Sensing Using Fluorescent Nanothermometers", ACS Nano **4**, 3254 (2010)

[3] K. Kashiwagi S. Yamashita, S.Y. Set, „In-situ monitoring of optical deposition of carbon nanotubes onto fiber end", Opt. Express **17**, 5711 (2009)

Scalable fabrication of twisted aperiodic multicore fibers for next-generation lens-less endoscopy

Ronja Stephan[1], Elias Scharf[2], Kinga Zolnacz[3], Katharina Hausmann[1], Matthias Ließmann[1], Lea Kötters[1,4], Jürgen Czarske[2], Detlev Ristau[1,4,5], Robert Kuschmierz[2], Michael Steinke[1,5]

1. Leibniz University Hannover, Institute of Quantum Optics, Welfengarten 1, 30167, Hannover, Germany

2. TU Dresden, Chair of Measurement and Sensor System Technique, Helmholtzstr. 18, 01069 Dresden, Germany

3. Wroclaw University of Science and Technology, Department of Optics and Photonics, Wyspianskiego 27, 50-370 Wroclaw, Poland

4. Cluster of Excellence PhoenixD (Photonics, Optics, and Engineering – Innovation Across Disciplines), Welfengarten 1A, 30167, Hannover, Germany

5. Leibniz University Hannover, QUEST-Leibniz-Research School, Callinstraße 36, D-30167 Hannover, Germany

Many biomedical applications require 3D imaging deep inside tissue. In this context, lens-less endoscopy, based on multicore fibers (MCFs), is a promising concept [1]. However, the cores in most conventional MCFs are arranged periodically. This periodicity causes artefacts during imaging in analogy to higher-order diffraction side lobes of periodic gratings. These in turn have a crucial impact on the performance, e.g. spatial resolution, of the endoscope. Such imaging artefacts can be reduced significantly by aperiodic core arrangements [2] but a scalable fabrication concept of such fibers is missing, so far. State-of-the-art aperiodic MCFs feature only insufficiently low core numbers of less than around 200. Furthermore, since lens-less endoscopy is sensitive to dynamic bending of the fiber, solutions like twisting [3] of the fiber must be incorporated into the fabrication. This enables correction of phase distortion by static DOEs produced by 2-Photon Polymerization on the fiber facet [4].
Our advanced and scalable fabrication approach is based on an iterative stack&draw approach as presented in [5]. However, we combined it with in-depth numerical studies in order to identify the ideal trade-off between sidelobe suppression, core density, and cross-talk. Simulations revealed that a randomized stacking of so-called subgroups with three off-centric cores is the most promising approach. With a core NA of around 0.3, two MCFs were fabricated so far: (i) a 250 µm fiber with 420 cores and (ii) a 333 µm fiber with 1281 cores, see Fig. 1 (a). The 1281 core fiber has a design core diameter of around 1.5 µm, which means the cores are theoretically single-mode for wavelengths larger than 470 nm, which was confirmed experimentally. Furthermore, to characterize the cross-talk in the fiber, experiments with a structured illumination of a 2 m long sample of the 1281 core fiber were performed. A rectangular 75 µm x 100 µm window of one fiber facet was illuminated by a standard microscope. Imaging of the other facet did not show any power leakage into surrounding cores, see Fig. 1 (b). In parallel to the fiber fabrication, two approaches were tested to twist the fibers: (i) rotation of the fiber preform during fiber drawing and (ii) post-production twisting of the MCF within a fiber processing station. While the first approach yielded a twist period of around 1 cm^{-1}, the second approach provided a period of up to 2.5 cm^{-1}, see Fig. 1 (c). Preliminary experiments regarding the cross-talk in the twisted fibers revealed that it is not substantially influenced by the twisting.
In our contribution, we will present a scalable fabrication method of aperiodic MCFs assisted by numerical optimization and the MCF characterization. In addition, we will discuss the mentioned experiments regarding twisting of the MCFs as well as the resulting imaging properties.

a) b) c)

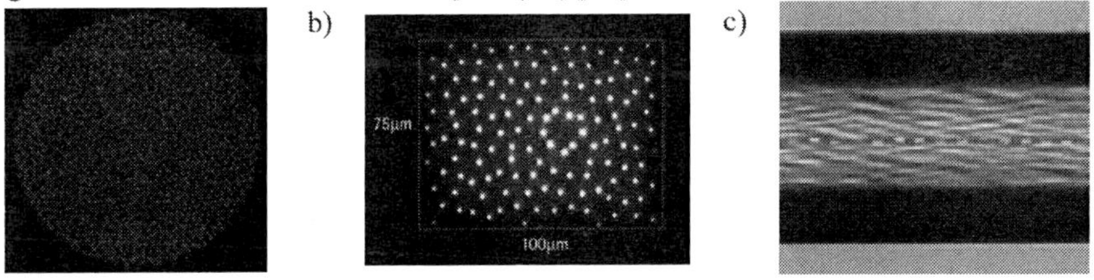

Fig. 1: *(a): The 1281 cores of the aperiodic multicore fiber (b): Structured illumination did not indicate any substantial cross-talk. (c): Side view of the twisted fiber with a twist period of 2.5 cm^{-1}.*

Literature

[1] R. Kuschmierz, E. Scharf, N. Koukourakis, and J. W. Czarske, "Self-calibration of lensless holographic endoscope using programmable guide stars," Optics Letters, 43(12), 2997–3000 (2018)
[2] S. Sivankutty, V. Tsvirkun, O. Vanvincq, G. Bouwmans, E. R. Andresen, and H. Rigneault, "Nonlinear imaging through a Fermat's golden spiral multicore fiber, " Optics Letters, 43(15), 3638–3641 (2018)
[3] V. Tsvirkun, S. Sivankutty, K. Baudelle, R.Habert, G. Bouwmans, O. Vanvincq, E. R. Andresen, and H. Rigneault, "Flexible lensless endoscope with a conformationally invariant multi-core fiber, " Optica 6, 1185–1189 (2019)
[4] R. Kuschmierz, E. Scharf, D. F. Ortegón-González, T. Glosemeyer, and J. W. Czarske, "Ultra-thin 3D lensless fiber endoscopy using diffractive optical elements and deep neural networks," Light: Advanced Manufacturing, 2(4): 415–424 (2021)
[5] S. Sivankutty, V. Tsvirkun, G. Bouwmans, D. Kogan, D. Oron, E. R. Andresen, and H. Rigneault, "Extended field-of-view in a lensless endoscope using an aperiodic multicore fiber," Optics Letters 41, 3531–3534 (2016)

EPJ Web of Conferences **267**, 02028 (2022) https://doi.org/10.1051/epjconf/202226702028
EUROPHOTON 2022

Record power transmission of intense 343 nm UV radiation in a single-mode inhibiting coupling hollow-core fiber exceeding 20W of 10-ns pulses

Florian Leroi[1] , Arnaud Guillossou[2] , Julien Didierjean[2], Julien Saby[2] , Johan Boullet[1]

[1] ALPhANOV, Institut d'Optique d'Aquitaine, Rue François Mitterrand, 33400 Talence, FRANCE;
[2]BLOOM Lasers, Cité de la Photonique, Bâtiment Electre, 11 Avenue de Canteranne, 33600 Pessac FRANCE.

Silica fibers are successful at delivering high-power high-energy signals in the near-infrared and visible but suffer from high absorption and color formation in the UV spectrum. Hollow-core negative-curvature anti-resonant fibers are promising alternatives as UV radiation has a low overlap with the guiding microstructure. This type of fibers is promising as pulsed UV radiation can be delivered with a few tenths of dB/m transmission loss[1,3]. However, their power scaling is notably hindered by the power handling of the thin capillary microstructure. Single-mode tubular anti-resonant fibers were reported delivering 106mW 20ps 355ns pulses with 66% coupling efficiency over 1m of fiber[1], 13.75mW 17ns 266nm pulses with 40% transmission over 2m of fiber[2], and ~180mW 350fs 343nm pulses with ~52% transmission over a 5 m long fiber[3]. We report here 2 orders of magnitude of gain on this state of the art, with on a record single-mode delivery of 23.3W (155µJ) with 89.1% transmission from a 343 nm, 10 ns, 150 kHz laser source. Fig. 1(left) depicts the experimental setup for this record UV beam fiber transport experiment. The used Flex-UV laser source developed by BLOOM Lasers delivers a 343nm single-mode signal (M^2=1.04) and was parametered to deliver a 10ns-long 150kHz diverging output. Two standard fused-silica plano-convex lenses of 300mm and 100mm focal length shaped the beam down to 19µm, the mode-field diameter of the fiber LP_{01} mode, see Fig. 3. The fiber was fabricated by the stack-and-draw technique. The 2m-long hollow core cladding fiber cladding is composed of 8 non-touching tubes with a diameter of 11 µm and a thickness of 600 nm. The tubes are arranged to form the surround of a hollow core with a diameter of 27 µm. The endface of the fiber is shown on the right of Figure. 1. The fiber power handling is also presented on Fig. 1. The delivery of up to 23.3W is reported with a 90.0% average combined coupling and transmission efficiencies but notably peaking at 95.6% at 5.63W. The fiber output mode was imaged onto a CMOS camera using a +50mm-focal-length lens, four tilted wedges and neutral density filters. The single-mode output maintains over the power range as reported in Fig. 1.

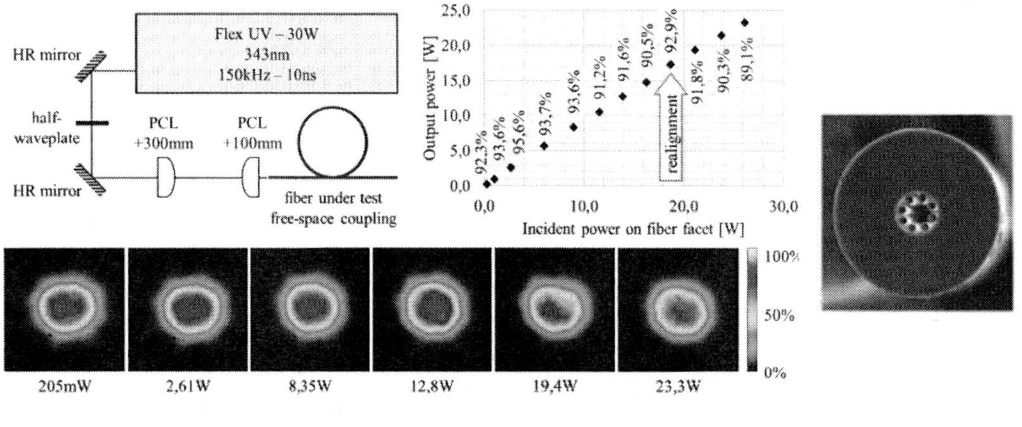

Figure. 3 : Power scaling: Schematic of the optical set-up; HR: High-Reflection; PCL: Plano-Convex Lens (top right), output against input with transmission coefficients (top right) with corresponding output beam profile from a CMOS camera (bottom).. Right : end face of the hollow core fiber

In conclusion, we have demonstrated thanks to the use of a state of the art high power UV fiber laser and an ultra-low loss IC-HCPCF, we managed to add 2 orders of magnitude on the power handling of fibers in the UV range, delivering more than 23 W of UV light at 343 nm on 2 m of IC-HCPCF.

References

[1] *Gao, S.-F., Wang, Y.-Y., Ding, W., and Wang, P., "Hollow-core negative-curvature fiber for UV guidance," Opt. Lett. 43, 1347-1350 (2018). [4] Yu, F., Cann, M., Brunton, A., Wadsworth, W., and Knight, J., "Single-mode solarization-free hollow-core fiber for ultraviolet pulse delivery," Opt. Express 26, 10879-10887 (2018)Lhermite, J., et al., "Millijoule Level Laser System Delivering 1 to 18 Ghz Intra-Train Repetition Rate Picosecond Pulses for Materials Processing,"* CLEO/Europe-EQEC, *2019, pp. 1-1.*

[2] *Yu, F., Cann, M., Brunton, A., Wadsworth, W., and Knight, J., "Single-mode solarization-free hollow-core fiber for ultraviolet pulse delivery," Opt. Express 26, 10879-10887 (2018).*

[3] *Michieletto, M., Jakobsen, C., Lyngsø, J. K., Oulianov, D., Triches, M., Olesen, A. S., Johansen, M. M., Patel, R. S., Murphy, T., Maack, M. D., "Visible and UV femtosecond pulse light delivery in anti-resonant hollow core fibers," Paper 11260-47, 1 February 2020 1:40 PM TO 3:40 PM | Part of SPIE*

Highly birefringent all-normal dispersion silica fiber with flat dispersion profile in the 1200–2100 nm wavelength range

Olga Szewczyk[1], Gabriela Statkiewicz-Barabach[2], Jacek Olszewski[2], Kinga Żołnacz[2], Mariusz Makara[3], Krzysztof Poturaj[3], Paweł Mergo[3], Jarosław Sotor[1], Grzegorz Soboń[1], Wacław Urbańczyk[2]

1. *Laser & Fiber Electronics Group, Wroclaw University of Science and Technology, 50-370 Wroclaw, Poland*
2. *Department of Optics and Photonics, Wroclaw University of Science and Technology, 50-370 Wroclaw, Poland*
3. *Laboratory of Optical Fibers Technology, Institute of Chemical Sciences, Maria Curie Sklodowska University, 20-031 Lublin, Poland*

Highly birefringent, polarization-maintaining (PM) optical fibers with all-normal dispersion (ANDi) are in great demand nowadays as a potential use for supercontinuum generation [1], dispersion management of ultrafast mode-locked oscillators [2] or as stretchers for chirped pulse amplification (CPA) systems [3]. Such optical fiber, suitable for abovementioned applications should have a high polarization extinction ratio (PER, of at least -20 dB), low transmission losses, and a dispersion profile with plateau over a possibly broad spectral range, covering the emission spectra of most common fiber lasers: Erbium- (1520 – 1600 nm) and Thulium-doped (1700 – 2100 nm). The dispersion compensating fibers (DCFs) available on the market are either non-PM, or designed to provide negative dispersion only in a narrow wavelength range. A broad and flat dispersion profile can be obtained in microstructured fibers with cladding containing a structure of air-holes [1]. However, splicing of such fibers is complex, lossy and, hence, less applicable for all-fiber systems. Here, we present a new design of an all-solid, polarization-maintaining ANDi fiber with a flat dispersion profile over the range of 1200 – 2100 nm, easily spliceable to conventional PM-Panda fibers.

The optical fiber was fabricated in the Laboratory of Optical Fibers Technology, Maria Curie Skłodowska University in Lublin, Poland. The dispersion plateau in normal range is obtained by high GeO_2 doping level in the core reaching 32%mol. The cross-sections of the three prototype fibers are presented in Fig.1 a-c together with the dimensions of the fiber cores. The outer diameters of the fibers were equal respectively to 125 μm, 122 μm and 127 μm. The fiber does not contain any air-hole structure which makes it easier for fusion splicing. The splicing loss can be optimised to the level of 0.28 dB for one splice, using conventional arc-fusion splicers. The fibers maintain a PER better than -22 dB after splicing to a standard PM Panda-type fiber. The dispersion profiles measured using spectral interferometry method are presented in Fig. 1 (d). The result directly indicates that the dispersion is flat over a broad range of 1200-2100 nm which makes it perfectly suitable for various applications including CPA, dispersion-management of ultrafast Er- and Tm-doped fiber oscillators, and broad coherent SC generation.

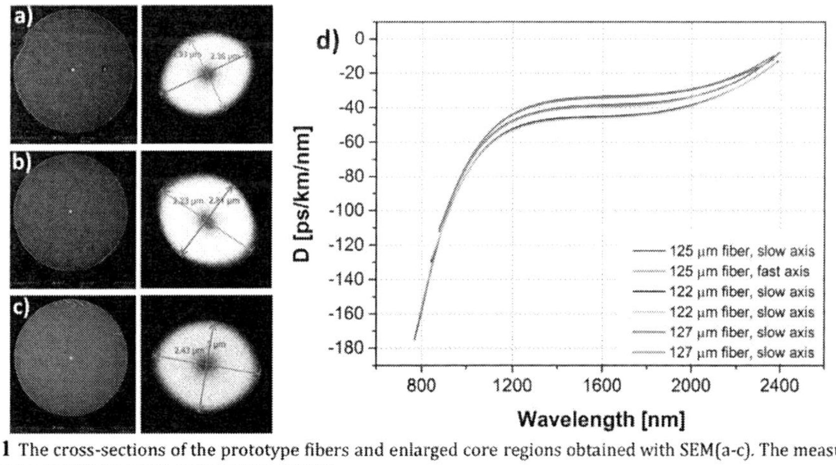

Fig. 1 The cross-sections of the prototype fibers and enlarged core regions obtained with SEM(a-c). The measured dispersion profiles for both polarization axis (d).

References

[1] A. Hartung, et al "Design of all-normal dispersion microstructured optical fibers for pulse-preserving supercontinuum generation," Opt. Express **19**, 7742–7749 (2011).
[2] J. Sotor, et al "All-polarization-maintaining, stretched-pulse Tm-doped fiber laser, mode-locked by a graphene saturable absorber," Opt. Lett. **42**, 1592-1595 (2017).
[3] I. Pavlov, et al "High-power high-repetition-rate single-mode Er-Yb-doped fiber laser system," Opt. Express **20**, 9471-9475 (2012).

Finite-size scaling behaviour in fully-connected equal-coupling multimode photonic networks

Oliver Melchert[1,2]

1. Leibniz Universität Hannover, Institute of Quantum Optics, Welfengarten 1, D-30167 Hannover, Germany

2. Cluster of Excellence PhoenixD (Photonics, Optics, and Engineering — Innovation Across Disciplines), Welfengarten 1A, D-30167 Hannover, Germany

The recently established connection between nonlinear multimode photonic systems and models of interacting classical spins [1], allows to now analyse dynamical equilibrium behaviour and collective phenomena in photonic systems using methods developed in statistical mechanics [2]. This unexpected synergy is highly valuable for studying effects that require a system-wide cooperation of all its degrees of freedom (DOF). In this regard we consider mean-field equal-coupling photonic networks [1], in which the elementary DOF are given by the complex-valued modes of the system. These modes are subject to a Kerr-type self-interaction and interact pairwise via uniform couplings. They can be viewed as photonic soft-spins: two-component spin vectors for which the mode-amplitude determines the length of the spin, see Figs. 1(a,b). Such a geometric picture exploits analogies to models of interacting classical spins, such as the XY model [2]. Previous studies of mean-field photonic networks unveiled [1], that, depending on the energy per mode, a second-order phase transition can be observed: if the average energy per mode is large, the system is in a disordered state in which the photonic spins are active and where there is no orientational order among the spins [Fig. 1(b)]; if the average energy per mode decreases, the system undergoes a continuous transition towards an ordered phase in which alignment forces induce global order [Fig. 1(a)].

Here, we perform long-time temporal evolution of the nonlinear equations of motion of the coupled modes using a stepsize controlled Runge-Kutta method of high order [3]. We address the issue of obtaining initial conditions for specified optical power and energy, for which we devised an effective optimization heuristic [4]. Based on extensive numerical simulations, we perform a comprehensive finite-size scaling analysis to estimate the critical points and critical exponents of the phase transition driven by the energy per mode. For this purpose we build upon the analogy to XY-type models by defining two-component and single-component order parameters, similar to those studied in statistical mechanics [2]. This allows to describe collective phenomena on equal-coupling photonic networks in terms of statistical summary measures that take the familiar form of the magnetization [Fig. 1(c)] and finite-size susceptibility [Fig. 1(d)] of statistical mechanics models. Particular attention is paid to the scaling behavior right at the critical point [Fig. 1(e)], at which we account for corrections to scaling by means of simple scaling laws. We take care to sample the statistical properties of the model in dynamical equilibrium and we compare our findings to exact results whenever possible.

In case of photonic networks, systems of interest naturally exhibit a finite number of DOF, hence finite-size effects as well as corrections to scaling can become an important feature whenever collective phenomena are addressed or systems with specific properties are to be designed. In this regard, the reported results are of fundamental interest in nonlinear optics.

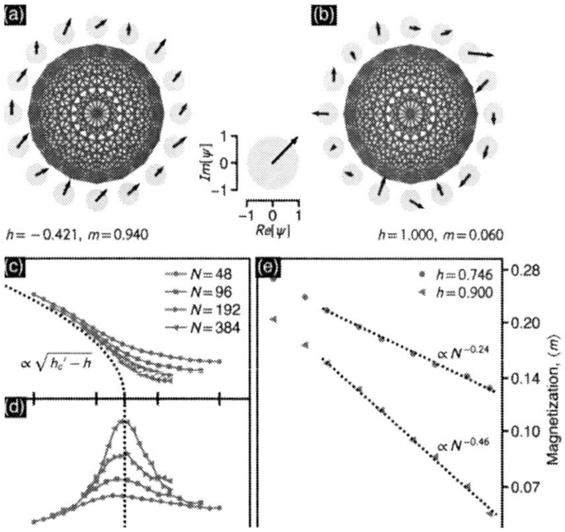

Fig. 1 Fully-connected equal-coupling photonic networks. Soft-spin configurations in (a) the ordered phase, (b) the disordered phase. (c) Scaling behaviour of the time-averaged magnetization, and, (d) scaling behaviour of the finite-size susceptibility as function of the energy per mode. (e) Finite-size scaling at fixed values of the energy per mode.

References

[1] A. Ramos, L. Fernández-Alcázar, T. Kottos, and B. Shapiro, "Optical Phase transitions in Photonic Networks: a Spin-System Formulation," Phys. Rev. X **10**, 031024 (2020).
[2] J. Cardy, *Scaling and Renormalization in Statistical Physics*, (Cambridge University Press, Cambridge, 1987).
[3] E. Hairer, S. P. Norsett, and G. Wanner, *Solving ordinary Differential Equations I. Nonstiff Problems*, (Springer Verlag, Berlin, 1993).
[4] O. Melchert, "Critical point phenomena and finite-size scaling in mean-field equal-coupling photonic networks," preprint, arXiv:2203.09216 (2022).

Photocathode Laser based on a 3 GHz Electro-Optical Comb Generator for the Ultrafast Electron Diffraction Facility REGAE

Christoph Mahnke[1], Chen Li[1], Henrik Tünnermann[1], Caterina Vidoli[1], Uwe Große-Wortmann[1], Christoph M. Heyl[1,2], Lutz Winkelmann[1] and Ingmar Hartl[1]

[1]*Deutsches Elektronen-Synchrotron DESY, Notkestraße 85, 22607 Hamburg, Germany*
[2]*Helmholtz-Institute Jena, Fröbelstieg 3, 07743 Jena, Germany*

The technique of Ultrafast electron diffraction (UED) combines the excitation of a probe using an optical laser pulse with electron diffraction, allowing the investigation of fast dynamics occurring in matter. It is especially suited for targets with low atomic number, and thus interesting for organic chemistry and biology.

We present recent results on laser system that is going to be used as a photocathode laser at the REGAE UED facility [1] at DESY. One challenge for UED is the so-called space charge, which occurs due to the mutual repulsion of the electrons and which limits the coherence length and the achievable spatial resolution required for biological samples. Our approach to overcome this issue is to distribute the 100 fC charge required to generate sufficient diffraction signal to a burst of about 4500 electron bunches, each containing only about 140 electrons. To still allow a µs time resolution, a very high intra-burst repetition rate is necessary. The currently developed photocathode laser is required to produce 1.5 µs bursts of UV (257 nm), 0.1 nJ picosecond pulses with 3 GHz intra-burst repetition rate and a burst repetition rate of 12 to 50 Hz.

The system is implemented as an electro-optical comb generator, based on polarization maintaining (PM) components. The schematic of our setup is shown in panel (a) of Fig. 1. Light at 1030 nm from a continuous wave (CW) seed laser is spectrally broadened by generating sidebands using three fiber-coupled phase modulators (PM), achieving about 2 nm bandwidth. An additional fiber-coupled intensity modulator (IM) is set to pick the segments with linear chirp, which are compressed to pulses of about 2 ps duration in an integrated spatial light modulator (SLM). The signal afterwards is amplified to the 100 W level using two Ytterbium fiber amplifier stages PA1 and PA2. Frequency conversion to the fourth harmonic at 257 nm is done in two stages, where second harmonic (SHG) is generated in a 15 mm noncritically phase matched LBO, followed by the fourth harmonic generation (FHG) a 5 mm CLBO crystal UV.

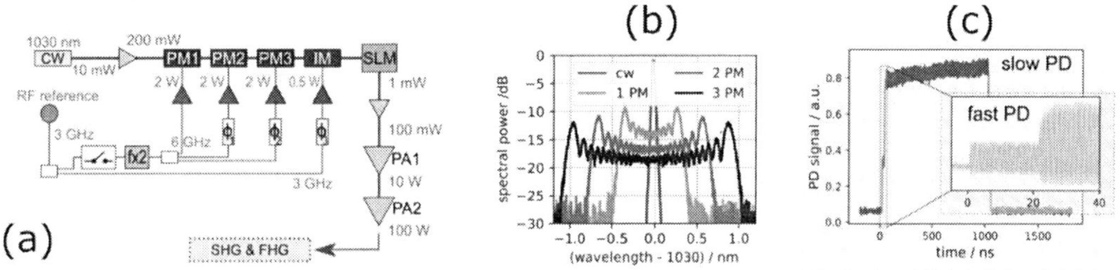

Fig. 1 (a) Schematic of the Photocathode laser system. (b) Optical spectra of the cw laser, and after each of the phase modulator stages. (c) Photo diode trace of a short burst generated by the system at 515 nm. The inset shows a zoom to the rising edge of the burst, resolving the pulse train of 3 GHz repetition rate.

As the radio-frequency (RF) fields driving the modulators are directly derived from the REGAE reference RF, the laser is inherently synchronized to the electron accelerator. To generate the required, µs-long bursts, an RF switch is used rapidly to turn the phase modulators on and off. Within durations of about 20 ns, the system thus can be switched between long pulse (~150 ps pulses) and short pulse (~1 ps) mode. Due to the strong peak-power dependency of the frequency conversion process, the required short and background-free UV bursts can be generated. We are currently in the progress of implementing the frequency conversion stages, reaching about 20 W of green light so far. Panel (c) of Figure 1 shows a photodiode trace of a microsecond long burst of the 515 nm (SHG) light; the inset a zoom to the rising edge of the 3 GHz burst. After completion and characterization of the system, the laser will be integrated in the REGAE facility in Q3 of 2022.

References

[1] M. Hada, J. Hirscht, D. Zhang, S. Manz, K. Pichugin, D. Mazurenko, S. Bayesteh, H. Delsim-Hashemi, K. Floettmann, M. Huening, S. Lederer, G. Moriena, C. Mueller, G. Sciaini, and D. Miller, "REGAE: New Source for Atomically Resolved Dynamics," in Research in Optical Sciences (OSA, 2012), p. JT2A.47.

[2] A. J. Metcalf, V. Torres-Company, D. E. Leaird, and A. M. Weiner, "High-Power Broadly Tunable Electrooptic Frequency Comb Generator," IEEE Journal of Selected Topics in Quantum Electronics 19(6), 231–236 (2013)

Enhanced Nonlinear Spectral Broadening in Multi-Pass Cells Using Molecular Gases

Moinuddin Kadiwala[1], Nazar Kovalenko[1], Kilian Fritsch[1], Semyon Goncharov[1], Oleg Pronin[1]

Helmut Schmidt University, Holstenhofweg 85, 22043 Hamburg, Germany

A nonlinear spectral broadening technique based on the phenomenon of self-phase modulation (SPM) using a waveguide-like approach, namely the multi-pass cells (MPC), has been shown to have excellent performance in numerous experiments [1]. One of the main benefits of such a geometry is simple and flexible dispersion management within the cell. Bulk or noble gases are typically used as nonlinear media for spectral broadening. Recently, molecular gases have been proposed as an alternative to noble gases in hollow-core fibers for enhanced and efficient spectral broadening [3,4]. In this work, we experimentally investigate the use of molecular gases in multipass cells and perform its direct comparison with the nonlinear spectral broadening in the noble gas. Such a realization promises to serve as a power-scalable, extremely broadband, simple and tunable source of ultrashort pulses. Such a source can facilitate many applications in ultrafast spectroscopy and microscopy [3, 4].

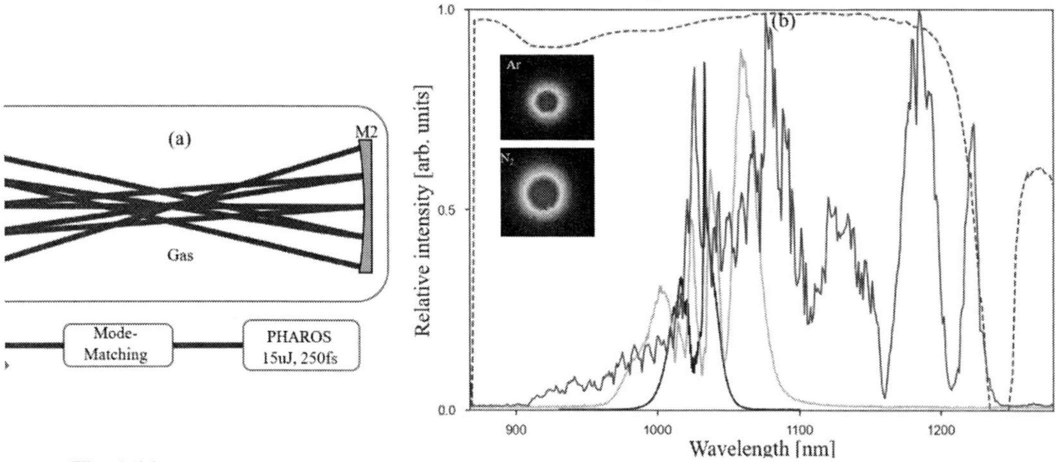

Fig. 1 (a) Schematic of a concave-concave multi-pass cell (MPC). (b) The output spectrum of Ar, N_2 at pulse energy of 15μJ, a repetition rate of 100 kHz and for N_2O at pulse energy of 15μJ and 10kHz repetition rate **inset**: output beam profile for Ar and N_2 .

We designed a MPC setup consisting of two concave mirrors with a 100 mm radius of curvature (ROC) introducing overall negative dispersion of -11.4 fs2 per roundtrip. The cell mirrors were separated by a distance of 130 mm along the optical axis. The eigenmode inside the MPC has a 1.3 mm and 0.20 mm $1/e^2$ radius on the mirrors M1 and M2, respectively. A commercial laser (Light conversion PHAROS) with 15 μJ pulse energy and 250 fs pulse duration was used as a driver for the experiment. The input beam was mode-matched to the eigenmode of the MPC using a single curved mirror with 1000 mm radius of curvature. Input pulses experienced 34 passes through the MPC. The MPC was placed in a gas-filled housing filled with 15.5 bars of Argon (Ar), Nitrogen (N_2), and Nitrous Oxide (N_2O). The overall transmission for Argon and Nitrogen was over 80 %, with clean output beam profile. However, the transmission dropped to less than 50% for N_2O.The output spectra were measured with a high resolution spectrometer (waveScan APE) (Fig 1(b)), resulting in a spectral span of 128 nm and 204 nm for Argon and Nitrogen, respectively. The broadening in N_2 and N_2O shows a strong red shift, Raman contribution, and is limited by the mirrors' bandwidth.

Conclusively, we demonstrated that it is possible to efficiently perform extreme spectral broadening in molecular gases by exploiting their effective rotational and vibrational nonlinearity. Nonlinear absorption and gas turbulences seem to be the main problems in operating such systems at high repetition rates and average power. Systematic investigation of these instabilities is a work in progress.

References

[1] Fritsch,K., Poetzlberger, M., Pervak, V., Brons, J., Pronin, O., "All-solid-state multipass spectral broadening to sub-20fs," Optics letters **43**, 4643-4646 (2018).
[2] Beetar, J. E., Nrisimhamurty, M., Truong, T. C., Nagar, G. C., Liu, Y., Nesper, J., ... & Chini, M., "Multioctave supercontinuum generation and frequency conversion based on rotational nonlinearity," Science Advances, *6*, eabb5375 (2020).
[3] Lefort, C., "A review of biomedical multiphoton microscopy and its laser sources," J. Phys. D: Appl. Phys. **50**, 423001 (2017).
[4] Wehbi, S., Mansuryan, T., Krupa, K., Fabert, M., Tonello, A., Zitelli, M., Ferraro, M., Mangini, F., Sun, Y., Vergnole, S., Kano, H., Wabnitz, S., Couderc V., "Continuous spatial self-cleaning in GRIN multimode fiber for self-referenced multiplex CARS imaging," Optics Express **30**, 16104 - 16114 (2022).

Adaptive liquid-core optical fibers for advanced soliton control

Mario Chemnitz[1], Ramona Scheibinger[1], Johannes Hofmann[1], Saher Junaid[1], Markus A. Schmidt[1,2]

1. Leibniz Institute of Photonic Technology, 07745 Jena, Germany

2. Otto-Schott Institute for Material Science, Friedrich-Schiller-University, 07743 Jena, Germany

Nonlinear optical functionalities in optical fibers are key for many applications in laser engineering, optical signal processing, and telecommunications. It appears to be a particularity of fibers that their linear and nonlinear optical properties are permanent after fabrication. Ultimately, this invariability inhibits to explore the manifold of operational regimes of all-fiber systems. Current advances in novel hybrid-material fibers may offer a solution to those limitations while opening the field of adaptive optics for the fiber technology community.

Here, we review our recent advances in both, (i) leveraging the fabrication particularities of liquid-core optical fibers (LCF) towards fiber-system integration and telecom compatibility, and (ii) utilizing the reconfigurable optical properties of such fibers via local temperature control, core composition, and higher-order mode excitation.

With their special optical properties, LCF unify the unprecedented nonlinear efficiency of soft-glass fibers and the extraordinary tunability and reconfigurability of gas-filled fibers (see Figure 1a). In particular liquids of low molecular complexity feature extraordinary, broadband transmission from the visible to the mid-IR, high nonlinearities, and outstanding coherence features of nonlinear processes such as supercontinuum generation [1]. Moreover, the large variety of liquids (e.g. carbon disulfide, chlorides, fluorides) and liquid compositions give access to vast dispersion landscapes based on simple, commercially available capillaries or micro-structured fiber types, providing an ideal testbed for dispersion-sensitive nonlinear experiments without the need for expensive waveguide fabrication. Yet, typical operation challenges, such as the use of external optofluidic mounts and free-space coupling, have long time prohibited LCFs to emerge outside the lab.

Here, we introduce the design principles and fabrication details of fiber-interconnectable, dispersion-optimized, step-index LCFs. We showcase the adaptive tuning potential of LCFs in three key applications: (1) controlling the soliton fission point [2], (2) wideband detuning of the soliton self-frequency shift at picojoule energy level (see Figure 1b) [3], and (3) tunable generation of femtosecond dispersive waves simultaneously at near- and short-wave infrared wavelengths [4].

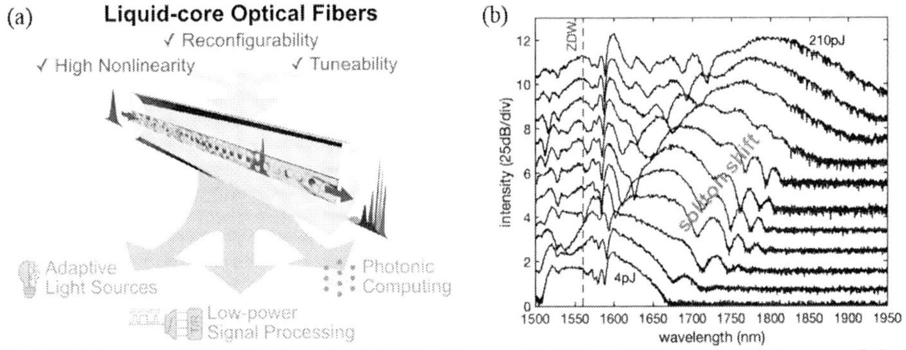

Fig. 1 (a) Advantages and application potential of liquid-core optical fibers. (b) Experimental showcase of pico-Joule-level soliton self-frequency shifting enabled by accurate design of the linear and nonlinear optical properties of straight-forward-to-fabricate step-index liquid-core fibers.

Our results emphasize liquid-core optical fibers as a promising platform featuring low coupling losses, low pump requirements, flexible dispersion landscapes, and hence, the ability to boost adaptive operation control and nonlinear functionalities in all-fiber-integrated systems. Those advanced may pave the way for next generation fiber technologies involving low-power, broadband signal processing, autonomous system optimization, as well as photonic machine learning.

References

[1] M. Chemnitz, M. Gebhardt., C. Gaida, F. Stutzki, J. Kobelke, J. Limpert, A. Tünnermann, & M.A. Schmidt, "Hybrid soliton dynamics in liquid-core fibres. *Nature Communications* **8**, 42 (2017).

[2] M. Chemnitz, R. Scheibinger, C. Gaida, M. Gebhardt, F. Stutzki, S. Pumpe, J. Kobelke, A. Tünnermann, J. Limpert, and M. A. Schmidt, "Thermodynamic control of soliton dynamics in liquid-core fibers," Optica **5**, 695 (2018).

[3] M. Chemnitz, S. Junaid, N. Walther, R. Scheibinger, K. Schaarschmidt, J. Kobelke, and M. A. Schmidt, "Tailoring soliton fission at telecom wavelengths using composite-liquid-core fibers," Opt Lett **45**, 2985 (2020).

[4] R. Scheibinger, N. M. Lüpken, M. Chemnitz, K. Schaarschmidt, J. Kobelke, C. Fallnich, and M. A. Schmidt, "Higher-order mode supercontinuum generation in dispersion-engineered liquid-core fibers," Sci. Rep. **11**, 5270 (2021).

EPJ Web of Conferences **267**, 02034 (2022) https://doi.org/10.1051/epjconf/202226702034
EUROPHOTON 2022

Packaging of an ultra-stable all-fiber-integrated NALM oscillator at 1 µm center wavelength for FEL facilities

Yi Hua[1], Henrik Tünnermann[1], Caterina Vidoli[1], Haydar Sarper Salman[1,2,3], Yuxuan Ma[1], Uwe Grosse-Wortmann[1], Lutz Winkelmann[1] and Ingmar Hartl[1]

1. *Deutsches Elektronen-Synchrotron DESY, Notkestrasse 85, Hamburg, 22603, Germany*
2. *Helmholtz-Institute Jena, Fröbelstieg 3, 07743 Jena, Germany*
3. *GSI Helmholtzzentrum für Schwerionenforschnung GmbH, Planckstrasse 1, 64291 Darmstadt, Germany*

In recent years, femtosecond fiber lasers mode-locked by nonlinear amplifying loop mirrors (NALM) [1] have shown many benefits for research and industrial applications such as high environmental stability, intrinsic low noise, and polarization-maintaining (PM) output. Previously we have demonstrated an all-fiber integrated, compact, and alignment-free NALM all PM Yb: fiber oscillator with sub-fs timing jitter [2]. Here we report on the next steps in packaging and engineering of this all-fiber compact oscillator. We developed a method to repeatably assemble lasers at a repetition rate required by DESY's FEL facilities.

All laser-oscillators at DESY's FEL facilities need to be synchronized with few-fs timing jitter to an integer divider of the accelerating RF field at 1.3 GHz. Since our oscillator is fully integrated and has no tuneable free-space optical components, it is a technical challenge to repeatably hit our target repetition rate of 1.3 GHz / 24 (54.167 MHz) within a few kHz. For this purpose, we developed the following method for cutting the intra-cavity fiber lengths such that the laser can be repeatably assembled to an overall cavity length within ~100 µm of the target: In the first step, we assemble a slightly long NALM oscillator cavity and measure the repetition frequency. Then, the laser is disconnected from its pump and connected to a low-coherence fiber reflectometer, a device which can determine a relative fiber length change with a resolution better than 60 µm. In the next step, we break the fiber loop by cleaving and measure the length difference between fiber-end B and a reference fiber (compare Fig 1a, left). The length of the fiber which needs to be cleaved off is then calculated and removed by a cleaver mounted on a micrometer resolution translation stage. The laser is then re-spliced. With this method, we achieve an accuracy of repeatability around ~100 µm, leading to the repetition rate accuracy of +/- 1.4 kHz for a 54.167 MHz laser cavity. The further coarse-tuning and repetition rate stabilization can be done by temperature control of the mounting plate for the intra-cavity fiber (10ppm length change or 0.5 kHz repetition rate change/ degree Celsius). Two piezo-actuators (slow and fast) are used for few-fs timing jitter synchronization.

Figure 1(b) shows the assembly of our all-fiber integrated NALM oscillator. Figure 1(c) shows the stabilization of the laser 24th harmonic of rep rate at 1.3GHz using only the resistive heater. As shown, a few kHz repetition rate change due to room temperature fluctuations can be easily compensated over long periods of time.

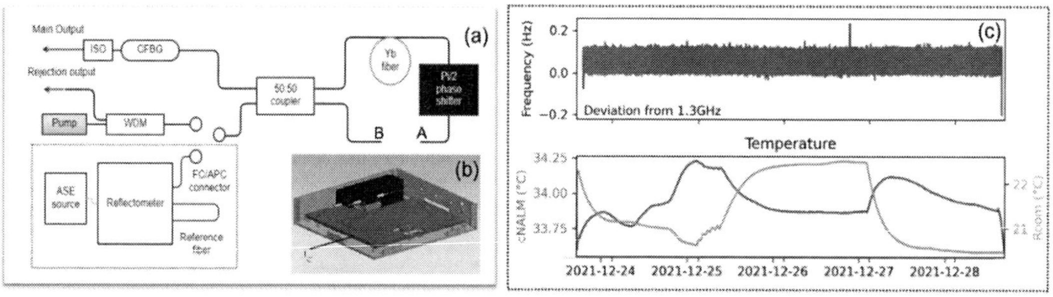

Fig. 1 (a) Setup configuration for precise fiber cutting; (b) oscillator electronic control box with the TEC plate for intra-cavity fiber; (c) Stabilization of 24th harmonic (1.3 GHz) of the oscillator rep rate by the TEC plate

In conclusion, we present the engineering and packaging of a compact all-fiber-integrated Yb: fiber oscillator with a precisely controlled repetition rate and without any tuneable free-space optical components. We developed a method to precisely cut the intracavity fiber lengths such that a target repetition rate can be reached repeatably within +/- 1.5 kHz. Further packaging and engineering steps, control system integration and long-term repetition rate stabilization with few fs timing jitter will be reported at the conference.

References

[1] W. Hänsel *et al.*, "All polarization-maintaining fiber laser architecture for robust femtosecond pulse generation," *Applied Physics B*, vol. **123**, No. 1, (2017).
[2] Y.Ma, et al. "Compact, all-PM fiber integrated and alignment-free ultrafast Yb: fiber NALM laser with sub-femtosecond timing jitter." Journal of Lightwave Technology 39.13 (2021): 4431-4438.

Stabilization of the unidirectionality phenomenon observed in a fully reciprocal fiber ring laser by retarding the seeding of Raman stokes

Muhammad Assad Arshad[1,2], **Dr. Alexander Hartung**[1], **Dr. Matthias Jäger**[1]
1. Leibniz-Institute for Photonic Technology, Albert Einstein Str. 9, 07745 Jena, Germany
2. Friedrich Schiller University, Faculty of Physics and Astronomy, Max-Wien-Platz 1, 07743 Jena, Germany

Recently a new form of unidirectionality was observed in a fully reciprocal fiber ring laser [1]. In contrast to the established forms of directionality induction e.g. by using components like circulators and isolators [2] or special cavity arrangements [3], this new form of unidirectionality requires no additional directional components. Moreover, unlike [4] the unidirectionality is observed at full laser operation far above the laser threshold. As no additional components are required, the new approach has the potential to circumvent any power and spectral constraints traditionally associated with directional devices, therefore making the approach suitable for high power and broad wavelength applications. Moreover, as it a previously unobserved phenomenon, there is potential for better understanding regarding the possible combination of known physical effects at play.

The setup in use is shown in Fig. 1a. It is a simple ring cavity including a pump diode, a power signal combiner (PSC), a Yb doped active fiber section, an output coupler (OC) or a WDM to couple some part of the light out and finally a km long section of passive Corning's HI-1060 fiber. As no directional components are involved the cavity initially starts its operations in the bidirectional regime where almost equal amount of power is emitted from the two counterpropagating directions. The situation changes as soon as pump power is increased, where above a certain threshold (named as switching power P_S) the output power from one (preferred) direction experiences a sudden increase while the opposite direction effective switches off and the cavity enters the unidirectional regime. In terms of the output spectrum (Fig. 1b), due to km long lengths of fiber involved the emitted spectrum is broadened due to stimulated Raman scattering (SRS), which further smoothens in the unidirectional regime. We have observed that the preferred direction in the unidirectional regime can be predefined with up to 100% certainty by managing the roundtrip losses and the P_S follows a power times length scaling [1], which is typical for thresholds associated with nonlinear phenomena like SRS. We therefore think that the underlying causes are nonlinear in nature.

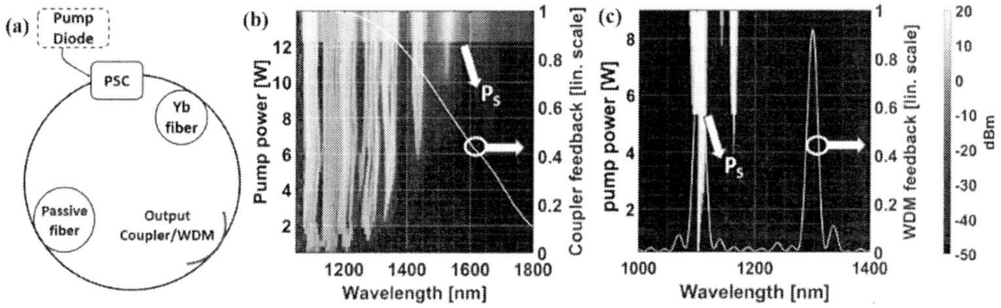

Fig. 1 a) Experimental setup. Excited spectrum of the preferred direction and feedback profile b) with a 0.01% output coupler (OC) and c) a WDM series suppressing SRS. P_s stands for the switching power at which cavity enters the unidirectional regime.

To investigate this further, we decided to inhibit the spectral broadening due to SRS by coupling the excited Stokes out of the cavity by using a series of WDMs in place of the OC. WDM series effectively act as the filter stage that hinders the seeding of Stokes by coupling them out of the ring. The results are shown in Fig. 1c), here we observe that the inhibition of spectral broadening results in more than 50% decrease in the P_s. Furthermore, limiting spectral broadening results in a consistent directional preference (counter clockwise) in the unidirectional regime. Moreover, the cavity also stabilizes in terms of P_s and output powers i.e. for multiple runs the P_s and output powers were highly reproducible. This is beneficial as this enables us to reduce cavity lengths considerably but still inducing unidirectionality at manageable pump powers, an essential step towards commercial realization of this phenomenon.

References

1. M. A. Arshad, A. Hartung, A. C. Pratiwi, and M. Jäger, "Observation of direction instability in a fiber ring laser," Sci. Rep. **11**(1), 4436 (2021).
2. D. Jalas, A. Petrov, M. Eich, W. Freude, S. Fan, Z. Yu, R. Baets, M. Popović, A. Melloni, J. D. Joannopoulos, M. Vanwolleghem, C. R. Doerr, and H. Renner, "What is — and what is not — an optical isolator," Nat. Photonics 7(8), 579–582 (2013).
3. S. Kharitonov and C.-S. Brès, "Unidirectional all-fiber thulium-doped laser based on theta cavity and fiber Bragg grating as filtering element," in *Lasers Congress 2016 (ASSL, LSC, LAC)* (OSA, 2016), p. AM5A.5.
4. W. J. Lai, P. Shum, and L. Binh, "Bidirectional optical bistability in a dual-pumped erbium doped fiber ring laser.," Opt. Express **12**(23), 5640–5 (2004).

Serrodyne optical frequency shifting using a nonlinear multi-pass cell

Henrik Tünnermann[1], Prannay Balla[1,2,3], Sarper H. Salman[1,2,3], Mingqi Fan[1,2,3], Skirmantas Alisauskas[1], Ingmar Hartl[1], Christoph M. Heyl[1,2,3]

1 Deutsches Elektronen-Synchrotron DESY, Notkestraße 85, 22607 Hamburg, Germany

2 Helmholtz-Institute Jena, Fr¨obelstieg 3, 07743 Jena, Germany

3 GSI Helmholtzzentrum f¨ur Schwerionenforschung GmbH, Planckstrasse 1, 64291 Darmstadt, Germany.

Ultrafast lasers are vital tools for a wide application range including strong-field physics, material processing precision spectroscopy and many more. Available laser platforms supporting femtosecond pulses at high peak and/or average power are primarily Ytterbium- (Yb), Ti:Sapphire- or Thulium-based. The center wavelength of such lasers is thus typically limited to very few operation points in the electromagnetic spectrum. Particularly, Yb-based lasers support femtosecond pulses at kW average powers [1] and, when complemented with suitable post-compression units, ultrashort pulse duration [2]. However, tuning high power femtosecond lasers to wavelength regions outside the amplifier gain bandwidth remains a major challenge, addressed so far at limited power levels using e.g. dispersive wave generation [3], soliton-shifting and Raman-based shifting techniques [4] or parametric frequency conversion.

Here we introduce an efficient and high-power compatible spectral shifting scheme based on a simple but very effective principle, Serrodyne frequency translation, as illustrated in 1(a) [5]. First, phase and optionally amplitude of an ultrashort laser pulse are shaped using a pulse shaper to produce a temporal saw-tooth pulse. In an optional second step, the shaped laser pulse can be amplified. The laser pulse is then spectrally shifted by acquiring linear phase ramp via self-phase-modulation within a dispersion-balanced multi pass cell (MPC). Afterwards, using a spectral filter, the wavelength-shifted spectrum is separated from residual broadband wavelength components.

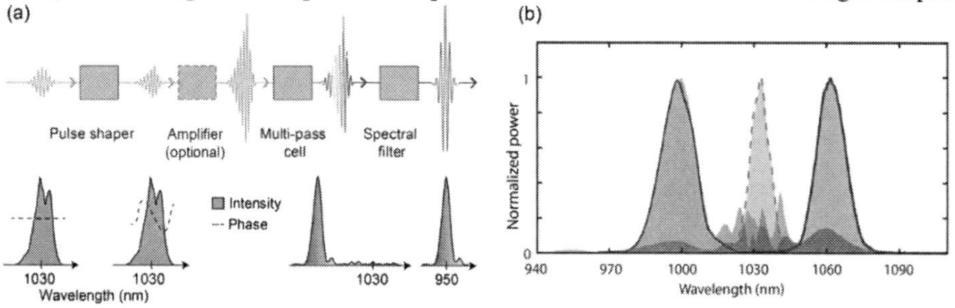

Fig. 1 (a) Schematic of the Serrodyne wavelength shifting scheme based on a dispersion-balanced MPC. (b) Experimental demonstration yielding Yb-laser pulses at high average power cantered around 1000 nm and 1060 nm.

We experimentally verify the principle by shifting the wavelength of an 80 W Yb frequency comb based on chirped-pulse amplification (CPA) delivering pulses with a duration of about 200 fs at 65 MHz repetition rate. Within the CPA chain, the stretched pulses are sent to a programmable phase shaper, followed by a two-stage amplification system and a compressor. The laser output pulses are then mode-matched into an MPC consisting of two concave dielectric mirrors (100 mm radius of curvature, negative mirror dispersion matched to about 10.5 mm fused silica (FS)). The MPC also contains multiple UV FS windows used as Kerr media. The throughput of the MPC is 82%, limited by the performance of the anti-reflection coatings of the windows. A saw-tooth-like pulses shape is applied for spectral shifting. The resulting strongly asymmetrically broadened MPC spectra are shown in 1(b). We use a dielectric dichroic mirror to filter the shifted spectral maxima located at 999 nm and 1062 nm, reaching an efficiency (MPC output to output behind the filter) of 63.1% and 66.7%, respectively.

Our results demonstrate a simple route for efficient spectral shifting of high-average and peak power lasers to wavelength regions not easily accessible by available ultrafast laser platforms, bringing ultrahigh peak and/or average power laser at flexible wavelength into reach.

References

[1] C. Grebing, M. M¨uller, J. Buldt, H. Stark, and J. Limpert, Opt. Lett. 45, 6250-6253 (2020).
[2] A.-L. Viotti, M. Seidel, E. Escoto, S. Rajhans, W. Leemans, I. Hartl and C. M. Heyl, Optica 449225 (2021).
[3] J. C. Travers,T. F. Grigorova, C. Brahms, F. Belli Nat. Photonics 13, 547-554 (2019).
[4] J. E. Beetar, M. Nrisimhamurty, T-C. Truong, et. al., Sci. Adv. 6, 34 (2020).
[5] P. Balla et al., "THz serrodyne optical frequency translator", submitted for publication (2022).

Hierarchical slaving for superior harmonic modelocking

***A. Şura*[1], *F. Ö. Ilday*[1,2]**

1. UNAM, Institute of Materials Science and Nanotechnology, Bilkent University, Ankara, 06800, Turkey

2. Department of Physics and Department of Electrical Engineering, Bilkent University, Ankara, 06800, Turkey

Despite its significance, control of multi-pulsing in modelocked lasers has had minimal success. We propose a conceptually new approach that self-organizes multiple pulses inside the cavity through a novel multi-scale interaction. The pulses in our laser are nonlinearly coupled through an acoustic effect but additionally through their nonlinear losses and the gain dynamics, resulting in a hierarchical interaction manifesting at the individual pulse level and collectively. We present an analytical model that conceptually extends the "slaving principle" of H. Haken for complicated dynamical systems to a hierarchical implementation over multiple levels, each distinguished by its timescale. While multi-timescale interactions are not new to modelocking, the hierarchical slaving perspective clarifies how to exploit such interactions for controlled evolution of the pulse pattern leading to harmonic modelocking (HML). In practice, we obtained reliable and stable HML with $\gtrsim$ 50 dB supermode suppression (SS) in a Mamyshev oscillator with variable filter offsets and *asymmetric* amplification arms (fig.1a). The pulse propagation is dominated by Kerr nonlinearity in one arm and similariton propagation in the other. All fibers are polarization-maintaining for environmental stability. Through the filter settings, we tune the pulse energy, which, equivalently, determines the number of pulses. The acoustic effect couples the energies and, consequently, the speeds to the temporal positions of the pulses, and this coupling allows us to control the pulse interactions by fine-tuning the filter settings. We achieved HML at every harmonic of our laser up to 1.2 GHz (83rd harmonic), limited by the average power (Fig.1). The output pulse duration was 150 fs, and the energy was 2 nJ at 1.2 GHz. To our knowledge, no fiber oscillator combines such a high repetition rate, strong SS, and short pulse duration [1, 2]. Moreover, these HML states are indefinitely stable. The slaving principle results in a simple dynamical model involving a 1D iterative map that describes the energy of each pulse [3]. By successively invoking the slaving principle to processes of higher timescales, we reach a differential system that describes the pulse positions, taking into account the acoustic effect and its coupling to the energies and speeds of the pulses. This model relates the nonlinear loss at the filters to the overall laser dynamics (fig.2) and clarifies how the pulse interactions can be controlled through the filter settings.

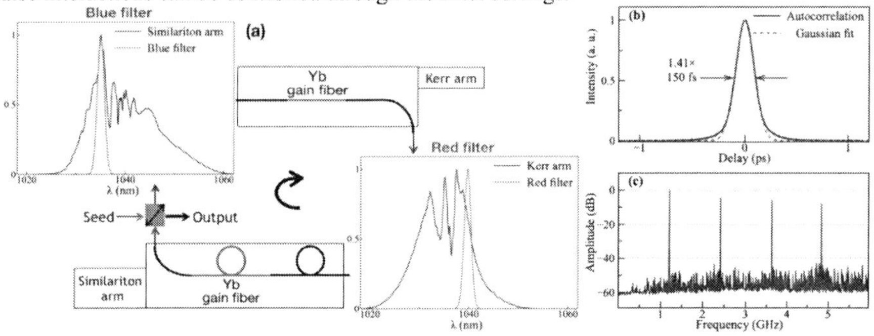

Fig. 1 (a) Schematic representation of our laser, showing the filters and the incident spectra for the 1.2 GHz state. (b) Autocorrelation of dechirped output pulses (similariton arm). (c) Radio Frequency spectrum showing ~50 dB SS

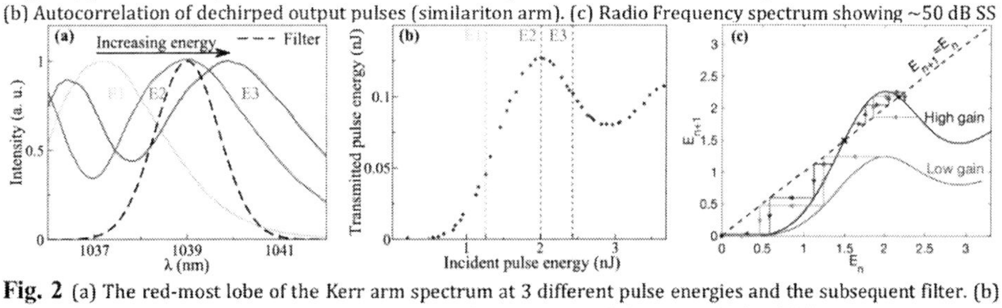

Fig. 2 (a) The red-most lobe of the Kerr arm spectrum at 3 different pulse energies and the subsequent filter. (b) The nonlinear transmisssion curve arising from the displacemmet of the spectral lobe. (c) Trajectories on typical energy maps. *n* denotes the roundtrip. The cross and dots indicate unstable and stable fixed points, respectively.

References

[1] G. Sobon, J. Sotor and K.M. Abramski, "*Sub-picosecond Graphene-based Harmonically Mode-Locked Fiber Laser With Repetition Rates up to 2.22 GHz,*" EPJ Web Conf **41**, 10001 (2013).

[2] J. Wang, X. Bu, R. Wang, L. Zhang, J. Zhu, H. Teng, H. Han, and Z. Wei, "*All-normal-dispersion passive harmonic mode-locking 220 fs ytterbium fiber laser,*" Appl. Opt. 53, 5088 (2014).

[3] F. Ö. Ilday in, *Theory and Ppractice of High-Energy Femtosecond Fiber Lasers*, (Cornell University, Ithaca, 2004), chapter 6.

[4] A. Şura, *Herarchically slaved multi-pulsing mode-lock dynamics*, (Bilkent University, Ankara, 2021).

EPJ Web of Conferences **267**, 02038 (2022)
EUROPHOTON 2022
https://doi.org/10.1051/epjconf/202226702038

Planar Polymer Optical Waveguide Coated with Metal-Organic Framework for CO_2 Sensing Application

Lei Zheng[1,2], Nils Keppler[2,3], Peter Behrens[2,3] and Bernhard Roth[1,2]

1. Hannover Centre for Optical Technologies, Leibniz University Hannover, Nienburger Straße 17, 30167 Hannover, Germany

2. Cluster of Excellence PhoenixD (Photonics, Optics, and Engineering-Innovation Across Disciplines), Welfengarten 1, 30167 Hannover, Germany

3. Institute of Inorganic Chemistry, Leibniz University Hannover, Callinstraße 9, 30167 Hannover, Germany

The need for the detection, sensing and monitoring of the greenhouse gas carbon dioxide (CO_2) for environmental protection has fostered the development of many different types of sensors with regard to the operation conditions and application fields [1,2]. A gas sensor works based on the active interaction between the sensing layer and the target gas. To date, gas sensors with high sensitivity and selectivity, compact size and low power consumption are increasingly demanded. Metal-organic frameworks (MOFs), which are porous crystalline materials that are assembled by the coordination of inorganic building units by organic linker molecules, exhibit great potential in advancement of novel gas sensors with improved performance due to (i) their high porosity, large surface area and tailor-made pore sizes, relevant for materials issue and (ii) the properties of minimal drift, high gas specificity to other gases, relevant for the optical gas sensing issue [3,4].

In this work, we propose a simple and resource- and cost-efficient gas sensor based on a planar polymer optical waveguide with a coated zeolitic imidazolate framework (ZIF)-8 thin film for CO_2 detection and sensing at the near infrared wavelength 850 nm. ZIF-8 has a large surface area of 1947 m^2/g, a narrow pore aperture diameter of 3.4 Å, as well as higher thermal and chemical stability than other MOFs, which enables its efficient adsorption of different gases [5]. A low-cost and efficient hot embossing technique was introduced for the fabrication of the proposed waveguide structure together with an easy-to-implement technique for the realization of the MOF film. A microscope image of the prepared structure is shown in Fig. 1a. Nitrogen (N_2) and CO_2 were used as the reference gas and the target sensing gas, respectively.

Experimental results show that the waveguide sensor exhibits rapid optical response and good sensitivity to pure CO_2 (see Fig. 1b) as well as CO_2 with different concentrations (see Fig. 1c). The sensing device also exhibits adsorption time of 6 s and desorption time of 16 s to CO_2, respectively. A sensitivity of approx. 2.5 μW/5vol% and limit of detection down to 5vol% were obtained. In addition, the sets of cycling experiments demonstrate the reversibility and reproducibility of the sensing performance of the proposed planar polymer waveguide sensor. This proof-of-concept study enables a more flexible fabrication of polymer waveguides, simplifies the introduction of MOF-based optical devices and thus opens up new application potentials of this material combination for selective gas detection and sensing.

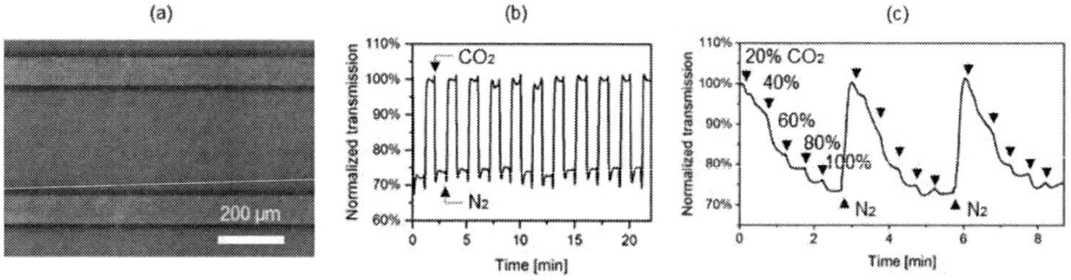

Fig. 1 (a) Microscope image of ZIF-8-coated polymer optical waveguide, which exhibits green color under optical microscope. (b) Dynamic response of ZIF-8-coated waveguide sensor to CO_2. (c) Dynamic response of ZIF-8-coated waveguide sensor to CO_2 with different concentrations.

References

[1] Z. Yunusa, M. N. Hamidon, A. Kaiser and Z. Awang. "Gas sensors: a review." Sens. Transducers 168, no. 4, 61-75 (2014).
[2] X. Liu, S. Cheng, H. Liu, S. Hu, D. Zhang and H. Ning. "A survey on gas sensing technology." Sensors 12, no. 7, 9635-9665 (2012).
[3] S. Yan, S. Zhang, W. Xie, L. Gai, H. Yuan, D. Zhang, H. Zhang, X. Liu, W. Yang and Z. Chi. "Chemiresistive ethanol sensors based on $In_2O_3/ZnSnO_3$ nanocubes." Sensors and Actuators Reports 4, 100099 (2022).
[4] J. Hodgkinson and R. P. Tatam. "Optical gas sensing: a review." Measurement Science and Technology 24, no. 1, 012004 (2012).
[5] K. S. Park, Z. Ni, A. P. Côté, J. Y. Choi, R. Huang, F. J. Uribe-Romo, H. K. Chae, M. O'Keeffe and O. M. Yaghi. "Exceptional chemical and thermal stability of zeolitic imidazolate frameworks." Proceedings of the National Academy of Sciences 103, no. 27, 10186-10191(2006).

EPJ Web of Conferences **267**, 02039 (2022) https://doi.org/10.1051/epjconf/202226702039
EUROPHOTON 2022

Four wave mixing in multimode hollow core waveguides with a two-color pump for the thorium nuclear clock

I. Babushkin[1,2], Ph. Mosel[1,2], K. S. Karda[1,2], A. Demircan[1,2], A. Trabattoni[1,3], M. Kovacev[1,2], U. Morgner[1,2]
1. Institute of Quantum Optics, Leibniz University Hannover, Welfengarten 1, 30167 Hannover, Germany
2. Cluster of Excellence PhoenixD (Photonics, Optics, and Engineering – Innovation Across Disciplines), Hannover, Germany
3. Center for Free-Electron Laser Science (CFEL), Deutsches Elektronen Synchrotron (DESY), 22607, Hamburg, Germany

Generation of high-energy pulses at wavelengths < 200 nm is a daunting task. One of the most important applications is a nuclear clock: the low-energy resonance of Th-229 is at around 8 eV, corresponding to $\sim$150 nm wavelength [1]. The clock, based on transitions in nuclei is less sensitive to perturbations and therefore more precise than atomic clocks. On the other hand, high isolation from the environment makes it difficult to excite, since high-intensity fields are required. Current methods to approach frequencies in this range involve typically nonlinear processes in gases, including high ionization-based harmonic generation (HHG) [2], resonant dispersive wave (RDW) from collapsing optical pulses [3] or four wave mixing (FWM) [4-8]. HHG and RDW are inherently bounded to short pulses. FWM in gases can potentially work both with short and long pulses but up to now also showed low efficiency and minor tunability. The primary FWM approach is cascaded FWM [4-8] using collinear phase-matching (PM) [4-6] or quasi-PM [7] in waveguides, or noncollinear PM in crossing filaments [8]. Using higher order modes (HOM) of waveguides was also reported for 3rd harmonic generation [9].

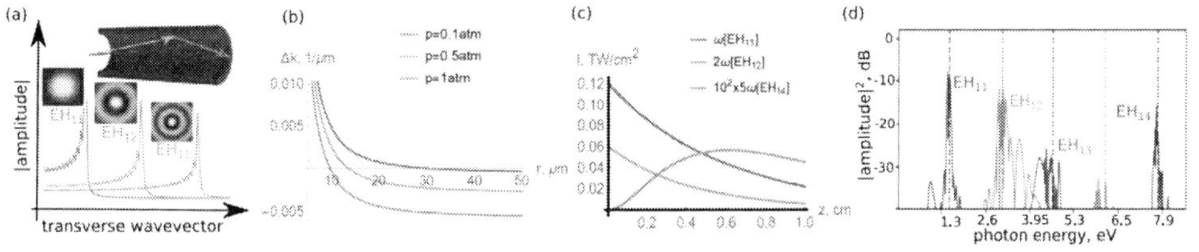

Fig. 1 (a) Higher-order modes in hollow-core waveguide for noncollinear PM. (b) PM in argon-filled capillary for different pressures and HOM given in (c). (c) Long-pulse FWM (efficiency $\approx 1\%$). (d) Short pulse FWM calculated using UPPE, efficiency $\approx 3\%$.

Here we propose to use two- color pump pulses and HOM of hollow-core waveguides. Using HOM is equivalent to noncollinear FWM (Fig. 1a), but allows increased interaction distance in comparison to crossing beams. In our approach, the pump is located at the fundamental (FH) frequency ω and its second harmonic (SH), giving rise to the FWM process $2\omega + 2\omega + \omega = 5\omega$. Every frequency is located at different transverse harmonic (see Fig. 1). Tuning SH $\omega \to \omega + \delta$ allows also to tune the signal wavelength. PM can be achieved by tuning the diameter of the capillary and the gas pressure (see Fig. 1c).

In the simulations of narrowband nanosecond Fourier-limited pulses we used the equations for the harmonic amplitudes similar to [7]. For short pulses, we simulated the multimode extension [10] of the so called unidirectional propagation equation (UPPE). Our results (Fig. 1) show that, despite of strong losses for higher order modes and the pulse reshaping (for short pulses), it is possible to achieve few per cent conversion efficiency using practically achievable pump pulse configuration.

As a conclusion, we demonstrated theoretically efficient generation of tunable VUV pulses via FWM involving higher order modes and two-color pulses. This method is especially attractive to generate VUV light for thorium clocks.

References
[1] K. Beeks et al, Nat. Rev. Phys. 3, 238 (2021).
[2] F. Krausz, and M. Ivanov, Rev. Mod. Phys. 81, 163 (2009).
[3] J. C. Travers et al, Nat. Photon. 13, 547 (2019).
[4] P. Tzankov et al, Opt. Express, 15, 6389 (2007).
[5] I. Babushkin, and J. Herrmann, Opt. Expr. 16, 17774 (2008).
[6] I. Babushkin et al, Opt. Lett. 33, 938 (2008).
[7] U. K. Sapaev et al, Opt. Expr. 20, 22753 (2012).
[8] M. M. Ghotbi et al, Opt. Lett. 35, 3492 (2010).
[9] F. Tani et al, JOSA B 31, 311 (2014).
[10] I. Babushkin, et al, Opt. Express 18, 9658 (2010).

 https://doi.org/10.1051/epjconf/202226702040

Femtosecond OPO pumped by a high power ytterbium rod-type fiber laser mode locked at harmonic repetition rates

Valerian FREYSZ[1,2], Eric FREYSZ[2]

[1]*Laboratoire Ondes et Matière d'Aquitaine (LOMA),UMR 5798, CNRS-Université Bordeaux, 33400 Talence, France*
[2]*ALPhANOV, Institut d'optique d'Aquitaine, Rue François Mitterrand, 33400 Talence, France*

Given their interest in multiple, most notably in spectroscopy, high-speed optical sampling, optical frequency comb, super-continuum generation, and others, high repetition rate (HRR) femtosecond lasers have attracted a lot of attention in the last decade. In rare-earth-doped fiber, the HRR can be achieved by controlling the nonlinear polarization rotation in long and heavily doped erbium or ytterbium fibers. However, it is usually difficult to tune the central wavelength of oscillators by providing pulses at an HRR. Further extension in the IR can be achieved through pumping an optical parametric oscillator (OPO) by amplifying the output of an HRR laser system. Compared with SHG and SFG, OPOs generate pulses that can be tuned on a broad spectral range, with limited pulse-to-pulse energy fluctuations and almost a TEM_{00} beam profile. Hence, one of the simplest and efficient ways to generate tunable HRR femtosecond pulses on a broad spectral range is to pump an OPO with a high-power laser oscillator, where one can adjust the PRR. We perform an experiment using an ytterbium rod-type fiber that provides more than 4.5W output power and tens of nJ per pulses, in which the PRR is controlled by nonlinear polarization rotation. The pulses delivered by this laser are further used to pump an OPO. The OPO is shown to work without any further adjustments. The femtosecond signal pulse delivered by this singly resonant OPO is tunable from 1450nm to 1700nm.

The setup is sketched in Fig. 1. The ring cavity is designed around a 0.75m long large-mode-area (LMA) double-clad ytterbium rod-type fiber. A 4nm interference filter centered at 1040nm with transmission limits the spectral bandwidth of the laser. A nonlinear and partially transmitting mirror is built by combining a quarter-wave plate (QWP) and half-wave plates (HWP1, HWP2) with an optical isolator, which takes advantage of nonlinear polarization evolution (NPE) within the LMA fiber. The pulses yielded by the laser are sent through a pair of chirped mirrors that compensate for the chirp induced by the propagation of the laser pulse in the amplifying fiber. The linear cavity of the OPO is singly resonant for the signal pulse. The OPO is built around a 3mm long MgO-PPLN and two plane and two concave mirrors with a radius curvature of 10cm. The cavity length for synchronous pumping at the 108MHz first repetition rate of the fiber laser is 1.39m.

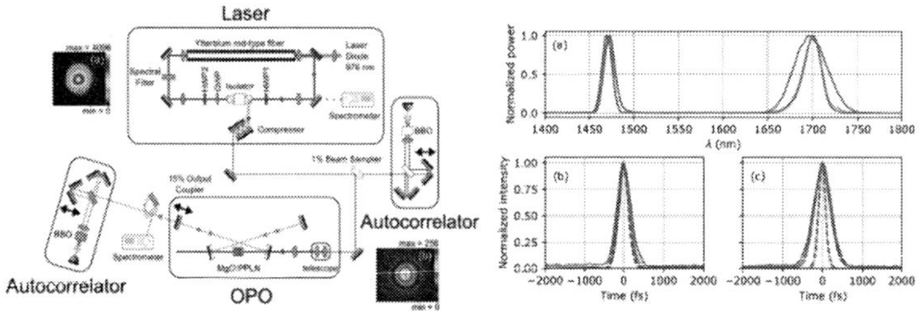

Fig. 1. (Left) Set-up of the femtosecond large-mode-area fiber laser and the linear optical parametric oscillator. QWP, quarter-wave plate; HWP, half-wave plate; MgO:PPLN, Magnesium doped periodically polarized lithium niobate. a-b) spatial profile of the beam delivered by the laser and OPO, respectively. (Right) Spectra and autocorrelation traces of the signal pulse delivered by the OPO. The curves are in green for the fundamental, in orange for the $2f_0$ and in blue for the $3f_0$ PRR. (a) 1470nm and 1695nm spectrum. FWHM for 1470nm is 14.0nm, 13.0nm and 13.0nm; FWHM for 1695 nm is 44.0nm, 24.0nm and 22.0nm. (b) Autocorrelation trace for the 1470nm spectrum. Measured durations (plain line) are 290fs, 360fs and 420fs; Fourier-limited durations (dot line) are 277fs, 347fs and 322fs for the 108MHz, 216MHz and 324MHz respectively. (c) Autocorrelation trace for the 1695 nm spectrum. Measured durations (plain line) are 490fs, 480fs and 500fs; Fourier-limited durations (dot line) are 181 fs, 190 fs and 212 fs for the 108 MHz, 216 MHz and 324 MHz PRR, respectively.

In conclusion, we have reported on an LMA rod-type ytterbium-doped all-normal dispersion fiber laser that can yield more than 4.5W output power and tens of nJ femtosecond pulses at up to 540 MHz. The control of the pulse repetition rate is based on NPR in a laser cavity. The compressed pulses delivered by this laser source are, without further amplification, used to pump an OPO. The femtosecond pulses delivered by this OPO were found to be tunable from 1400nm to 1700nm, and the average output power was higher than 1watt. More details about side mode suppression ratio, slope efficiency, RF spectra, spectra and AC traces at different repetition rates will be presented.

 https://doi.org/10.1051/epjconf/202226702041

Recent progress on rare earth amplifiers and lasers directly on silicon

Jonathan D. B. Bradley, Khadijeh Miarabbas Kiani, Henry C. Frankis, Cameron M. Naraine, Dawson B. Bonneville, Hamidu M. Mbonde, and Andrew P. Knights
Department of Engineering Physics, McMaster University, 1280 Main Street West, Hamilton, ON L8S 4L7, Canada

Silicon has become the platform of choice for low-cost, energy efficient and high-speed photonic microsystems for applications ranging from communications to healthcare. However, the development of on-chip amplifiers and lasers has been one of the persistent challenges in silicon photonics. Silicon itself is an inefficient light emitting material, which has led to various approaches to integrate different gain media on Si [1]. Much of these efforts have focused on heterogeneous or monolithic integration of III-V semiconductors leading to effective electrically pumped commercial devices [2]. However, optically pumped rare earth gain media can be considered an attractive alternative for many emerging applications, motivated by their low-cost, monolithic wafer scale deposition methods, broadband emission at near-infrared wavelengths of interest, and potential for low noise amplification and narrow-linewidth lasing [3]. However, integrating rare-earth amplifiers and lasers directly on silicon has proven challenging for several reasons, including the high confinement and relatively high loss in silicon waveguides making it difficult to achieve net optical gain, and incompatibilities with silicon photonic manufacturing (e.g., due to high temperatures required for deposition of rare-earth-doped gain materials).

Here we report on our recent developments on rare-earth-based gain and lasing directly on silicon photonic chips. We demonstrate optical gain and on-chip lasing around 1.8—1.9 μm in thulium-doped tellurium oxide (TeO_2:Tm^{3+}) coated silicon hybrid microdisks [4]. The silicon photonic chips are fabricated using a standard silicon foundry process, while the TeO_2:Tm^{3+} film is deposited at room temperature using a straightforward reactive RF magnetron co-sputtering post-processing step. We show lasing at 1.9 μm with a double-sided on-chip output power of > 1 mW, threshold launched pump power of 16 mW, and slope efficiency vs. absorbed pump power of 60% in point-coupled microdisks (Fig. 1a). We also demonstrate net optical gain within the hybrid TeO_2:Tm^{3+}-Si bus waveguide. In addition, we show sub-milliwatt threshold lasing in pulley-coupled microdisks which are engineered for stronger pump and reduced signal coupling strength (Fig. 1b). These results demonstrate the possibility of wafer-scale integration of thulium amplifiers and lasers in silicon photonic circuits for emerging applications near 2 μm. In addition, they provide motivation for exploring other dielectric gain materials, dopants and emission wavelengths (e.g., Er and 1.5 μm), and the development of novel monolithic devices and circuits (e.g., amplified silicon delay lines and tunable rare earth lasers) for applications in communications, computing, and sensing.

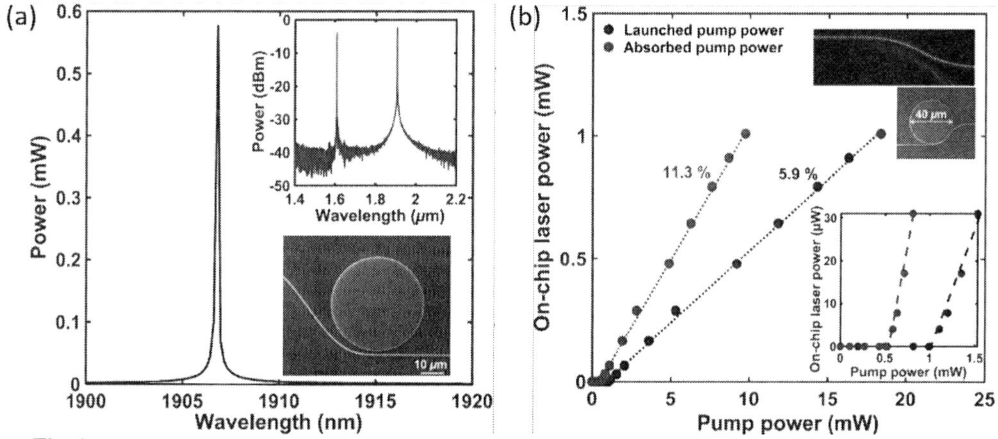

Fig. 1 (a) On-chip laser emission at 1.9 μm demonstrated in a hybrid TeO_2:Tm^{3+}-silicon microdisk (insets: broader spectrum showing residual 1.6-μm pump and laser signal light collected at chip output and top-view SEM of microdisk laser). (b) Sub-milliwatt-threshold lasing using a pulley-coupled microdisk design (insets: simulation of pulley coupled bus waveguide to engineer pump and signal coupling to and from the microdisk, respectively, SEM of pulley-coupled microdisk laser, and zoomed-in view of the laser curve near threshold).

References

[1] Z. Zhou, B. Yin, and J. Michel, "On-chip light sources for silicon photonics," Light Sci. Appl. **4**, e358 (2015).
[2] M. Tang, J.-S. Park, Z. Wang, S. Chen, P. Jurczak, A. Seeds, and H. Liu, "Integration of III-V lasers on Si for Si photonics," Prog. Quantum Electron. **66**, 1 (2019).
[3] N. Li, M. Xin, Z. Su, E. S. Magden, N. Singh, J. Notaros, E. Timurdogan, P. Purnawirman, J. D. B. Bradley, and M. R. Watts, "A silicon photonic data link with a monolithic erbium-doped laser," Sci. Rep. **10**, 1114 (2020).
[4] K. Miarabbas Kiani, H. C. Frankis, C. M. Naraine, D. B. Bonneville, A. P. Knights, and J. D. B. Bradley, "Lasing in a hybrid rare-earth silicon microdisk," Laser Photonics Rev. **16**, 2100348 (2022).

EPJ Web of Conferences **267**, 02042 (2022) https://doi.org/10.1051/epjconf/202226702042
EUROPHOTON 2022

Spectral Two-Photon Quantum Interference via Electro-optic Modulation Between Light States Of Different Photon Statistics

Anahita Khodadad Kashi[1, 2] and Michael Kues[1, 2]

[1]*Institute of Photonics, Leibniz University Hannover, Nienburger Str. 17, 30167 Hannover, Germany*
[2]*Cluster of Excellence PhoenixD, Leibniz University Hannover, Hannover, Germany*

Abstract: Frequency-domain two-photon quantum interference between a thermal field and a heralded-state is studied theoretically and experimentally, revealing the dependency of visibility on the multiphoton components within the heralded-state.

Photonic frequency-encoded quantum information processing can pave the way for the realization of global-scale quantum networks [1], integral to which is the scalable implementation of the Hong-Ou-Mandel (HOM) effect between independent light states of different photon statistics [2, 3]. In this work we demonstrate the first-time experimental implementation of the spectral HOM effect between independent single photons as well as the quantum interference between a thermal and a heralded state. To this end, an intrinsically stable and reconfigurable frequency processing circuit was used, which consisted of a mode-locked pulsed laser, two programmable filters, an electro-optic phase modulator (EOPM) – acting as a frequency counterpart of a conventional spatial domain beam splitter – and four superconducting nanowire single photon detectors (SNSPD) (see Figure 1a). The synchronized collection of coincidence events with the pulsed laser repletion-rate allowed for on-the-fly acquisition of the HOM effect. For the HOM effect between single photon states we measured a visibility of 74.31%±3.57% (see Figure 1 b), which falls well beyond the classical limit of 50%, offering an appropriate candidate for quantum communication and information applications. In the case of the HOM effect between a thermal field and a heralded state, we derived a theoretical relation, $V_{theory} = 1/(1 + (\overline{n}_{1,th}^2 + \overline{n}_2^2 + \overline{n}_{1,th})/(\overline{n}_{1,th}\,\overline{n}_2 + \overline{n}_2))$, describing the dependency of the visibility on the fields' thermal mean photon numbers per pulse period, i.e. $\overline{n}_{1,th}$ and $\overline{n}_2$. Specifically, this demonstrates that multiphoton components in the heralded state get engaged in quantum interference. Experimentally, by assuming the absence of the HOM effect at zero delay, we defined a reference point (see Figure 1c), revealing a visibility of $V_{experiment} = 43.20\% \pm 4.28\%$, which is in good agreement with the predicted theoretical value $V_{theory} = 47.30\% \pm 0.66\%$. Our results enable the implementation of important concepts such as quantum teleportation, entanglement swapping and complex states in large-scale frequency-domain quantum circuits, and thus provides a key element for future frequency-multiplexed quantum networks.

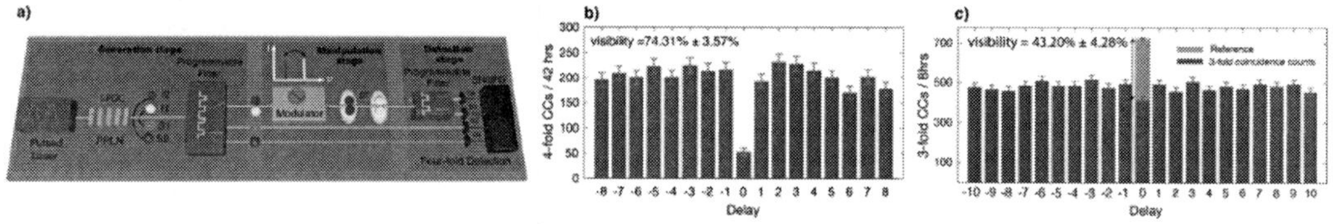

Figure 1. a) Experimental setup employed for HOM effect between two single photon states. (The setup for HOM between a thermal field and heralded state is not shown here.) **b)** Four-fold and **c)** three-fold coincidence counts as function of delay expressed as multiples of the pulse period.

References

[1] Kues, M. et al. Nat. Photonics 13 (2019), 170–179.
[2] A. Khodadad Kashi, M. Kues, Laser & Photonics Reviews 15 (2021), 2000464.
[3] X. Li, L. Yang, L. Cui, Z. Y. Ou, and D. Yu, Opt. Express 16 (2008), 12505.

Hybrid Integrated Nonlinear Photonics: From Chipscale frequency combs to cryogenic interconnects

Tobias J. Kippenberg (EFPL, Switzerland)

The development of optical frequency combs[1], and notably self-referencing, has revolutionized precision measurements over the past decade, and enabled counting of the cycles of light. Frequency combs, have enabled dramatic advances in timekeeping, metrology and spectroscopy. In 2007, it was discovered that such combs can also be generated using an optical microresonator[2] using parametric frequency conversion. Kerr combs also enable to generate dissipative temporal solitons (DKS)[3,4], which are formally solutions to a driven dissipative nonlinear Schrödinger equation, termed Lugiato-Lefever equation – first derived to describe spatial self-organization phenomena[5]. DKS have unlocked the full potential of Kerr combs enabling a deterministic route to broadband, and coherent optical frequency combs, whose bandwidth can be enhanced using soliton broadening phenomena, such as Soliton Cherenkov Radiation[6]. Such Solitons Kerr combs on a chip have enabled to realize counting of the cycles of light, realize dual comb spectrometers on a chip, enabled dual comb based ultrafast ranging[7], massively parallel coherent communication[8], and offered a novel approach to massively parallel FCMW LiDAR[9]. Recent advances in developing ultra low loss integrated photonics[10] based on silicon nitride (Si_3N_4), have enabled ultra-low propagation losses, enabling the direct integration with on chip pump lasers[11]. On the fundamental side, new and theoretically not previously predicted dynamics has been observed ranging from formation of soliton crystals[12], soliton switching[13], and new type of breather solitons[14], and emergent nonlinear dynamics in arrays of coupled resonators[15]. Nonlinear driven integrated photonics circuits are thereby providing a highly fruitful new playground for fundamental nonlinear science and applications alike. Beyond this, ultra-low loss integrated photonics are giving rise to novel applications: for realizing traveling wave optical parametric amplifiers to heterogeneous integration with Lithium Niobate to create ultra low voltage modulators for that can serve as cryogenic interconnects for superconducting quantum computing.

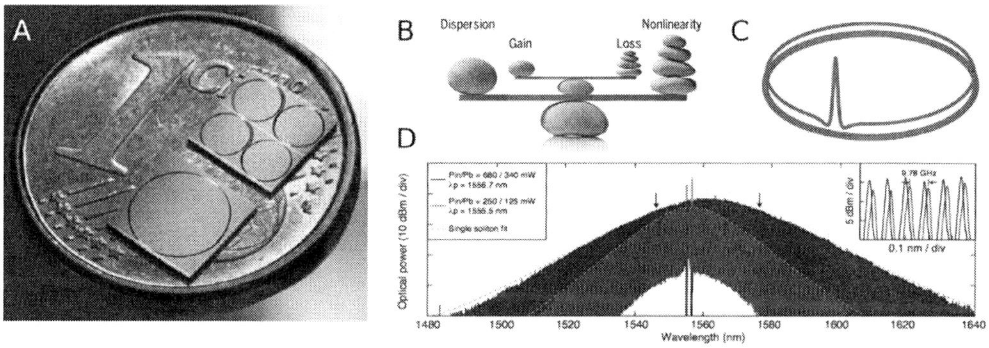

References

1. Udem, T., Holzwarth, R. & Hänsch, T. W. Optical frequency metrology. *Nature* **416**, 233–237 (2002).
2. Del'Haye, P. *et al.* Optical frequency comb generation from a monolithic microresonator. *Nature* **450**, 1214–1217 (2007).
3. Herr, T. *et al.* Temporal solitons in optical microresonators. *Nat Photonics* **8**, 145–152 (2014).
4. Kippenberg, T. J., Gaeta, A. L., Lipson, M. & Gorodetsky, M. L. Dissipative Kerr solitons in optical microresonators. *Science* **361**, eaan8083 (2018).
5. Lugiato, L. A. & Lefever, R. Spatial Dissipative Structures in Passive Optical Systems. *Phys. Rev. Lett.* **58**, 2209–2211 (1987).
6. Brasch, V. *et al.* Photonic chip–based optical frequency comb using soliton Cherenkov radiation. *Science* aad4811 (2015) doi:10.1126/science.aad4811.
7. Trocha, P. *et al.* Ultrafast optical ranging using microresonator soliton frequency combs. *Science* **359**, 887–891 (2018).
8. Marin-Palomo, P. *et al.* Microresonator-based solitons for massively parallel coherent optical communications. *Nature* **546**, 274–279 (2017).
9. Riemensberger, J. *et al.* Massively parallel coherent laser ranging using a soliton microcomb. *Nature* **581**, 164–170 (2020).

10. Pfeiffer, M. H. P. *et al.* Photonic Damascene process for integrated high-Q microresonator based nonlinear photonics. *Optica* **3**, 20 (2016).
11. Raja, A. S. *et al.* Electrically pumped photonic integrated soliton microcomb. *Nat. Commun.* **10**, (2019).
12. Cole, D. C., Lamb, E. S., Del'Haye, P., Diddams, S. A. & Papp, S. B. Soliton crystals in Kerr resonators. *ArXiv161000080 Phys.* (2016).
13. Guo, H. *et al.* Universal dynamics and deterministic switching of dissipative Kerr solitons in optical microresonators. *Nat. Phys.* **advance online publication**, (2016).
14. Guo, H. *et al.* Inter-mode breather solitons in optical microresonators. *ArXiv170505003 Phys.* (2017).
15. Tikan, A. *et al.* Emergent Nonlinear Phenomena in a Driven Dissipative Photonic Dimer. *ArXiv200506470 Nlin Physicsphysics* (2020).

Fiber based high power low noise single frequency lasers and applications

Clément Dixneuf [1,2], Dia Darwich[1], Roopa Prakash[1], Yves-Vincent Bardin[2], Mathieu Goeppner[2], Germain Guiraud[2], Nicholas Traynor[2], Adèle Hilico[1], Giorgio Santarelli[1]

[1]*Laboratoire Photonique Numérique et Nanosciences (LP2N), UMR 5298, CNRS-IOGS-Université Bordeaux, 33400 Talence, France*
[2] *Azur Light Systems, 11 Avenue de Canteranne, 33600 Pessac, France*

With the advances in fundamental science such as gravitational wave detection (GWD) [1], cold atom physics [2] and quantum computing [3], the need for single frequency (SF) high-power fiber lasers has been increasing over the last decade. The ability to maintain an excellent modal quality, the wide tuning range in the Ytterbium (Yb) band, the linear polarization and the optical low-noise operation is driving this technology as an excellent replacement for more established high-power bulk systems in the 1 µm region. We will present several very low noise high power laser sources at different wavelengths and the potential applications.
Although very high-power fiber lasers have been demonstrated in a master oscillator power amplifier (MOPA) configuration, SF operation faces some physical limitations, stimulated Brillouin scattering (SBS) being the principal one. This well-known non-linear effect has an impact on amplifier intensity noise and will eventually lead to catastrophic system damage at high enough power.
Several groups have already demonstrated SF operation up to hundreds of watts of continuous power. However, very few of these studies have focused on the noise characteristics of the amplifiers or the long-term stability of these systems. In this framework we have demonstrated a robust and efficient 365W MOPA laser with low intensity noise laser with excellent modal quality based on LMA fiber without SBS management (see Fig. 1 (a)). We have implemented full wide bandwidth intensity control able to pin down intensity noise close to 160dBc/Hz. The scaled down (130W) version of this laser is used in the Virgo GWD, with more than 5000h of operation. Beyond GWD this type of high power laser find applications in the generation of optical lattices and traps for modern atomic physics.

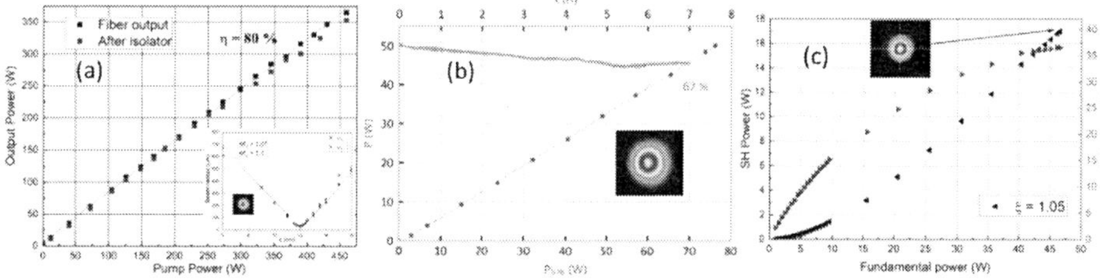

Fig. 1. (a): high power 1064nm MOPA, output power versus pump power at the fiber output (black) and after the isolator (red), M^2 (inset)(b): 1013 nm laser power vs pump power and slope efficiency (blue), 7-hour stability (orange) and mode profile (inset) (c) 1064/532 Second harmonic single pass generation SH power (black) and efficiency (red) versus fundamental power and mode profile (inset).

Another interesting case laying in the lower side of the Yb doped fiber operation window is 1013nm. This emission wavelength is critical for modern quantum processing units using neutral Rb atoms where tens of Watts are required, and no other available technology can compete with Yb fibers. The emission at such short wavelength is quite challenging due the strong signal re-absorption. This has two perverse effects: first, ASE around 1030 nm sees a much higher gain triggering parasitic lasing. Second, a strong pump intensity is needed to prevent signal reabsorption. The solution is to use short fibers, to limit ASE gain and ensure strong pump intensity all along the fiber but this impairs the efficiency and leaves a strong residual pump. Despite those problems we have demonstrated an efficient laser (67%) around 1013nm able to deliver up to 50W before photodarkening with low noise and good modal quality [5] (see fig 1 (b)). The industrial version of this lasers (10W) is currently used in the quantum processing unit under development in several laboratories and start-ups. By using Second Harmonic Generation (SHG) is also possible to produce high power visible sources in the green domain starting from Yb doped fiber laser. Again, low noise sources are necessary in modern physics, laser pumping and industrial applications. We have been able to demonstrate up to 17 W of low noise green radiation at 532nm by efficient single pass doubling (see Fig 1 (c)) and develop a model of the SHG intensity noise transfer in the saturation regime [6]

References

[1] LIGO&Virgo Scien. Coll., "Observation of Gravitational Waves from a Binary Black Hole Merger," Phys. Rev. Lett. 116, 061102 (2016).
[2] A. Mazurenko et al. "Implementation of a stable, high-power optical lattice for quantum gas microscopy", Review of Scientific Instr. 90, 033101 (2019)
[3] H. Levine et al. "High-Fidelity Control and Entanglement of Rydberg-Atom Qubits", Phys. Rev. Lett.121,123603 (2018)
[4] C. Dixneuf et al. "Ultra-low intensity noise, all fiber 365 W linearly polarized single frequency laser at 1064 nm," Opt. Express 28, 10960-10969 (2020)
[5] B. Gouhier et al. "Low-noise single-frequency 50 W fiber laser operating at 1013 nm", Laser Physics Letters, 16, 045103 (2019)
[6] C. Dixneuf et al. "Robust 17 W single-pass second-harmonic-generation at 532 nm and relative-intensity-noise investigation," Opt. Lett. 46, 408-411 (2021)

https://doi.org/10.1051/epjconf/202226702045

Phase locking of fiber laser array using quasi-reinforcement learning, principle and experiments

Vincent Kermene[1], Alexandre Boju[1-2], Maksym Shpakovytch[1], Geoffrey Maulion[1], Paul Armand[1], Alain Barthelemy[1], Agnès Desfarges-Berthelemot[1]

[1] Institut de recherche XLIM, Université de Limoges-CNRS UMR n°7252, Faculté des Sciences et Techniques, 123 ave. A. Thomas, F-87060 Limoges, France

[2] CILAS Ariane Group, 8 avenue Buffon, CS16319, F-45063 Orléans cedex2, France

Artificial intelligence is becoming more and more widespread in technological and scientific fields. Since 2019, it has extended to the coherent beam combining of laser array (CBC) [1], most of the time based on numerical studies. CBC is a key technique to reach high average power or high-energy short pulses while maintaining laser beam quality. In addition, the tiled-aperture arrangement of laser beam array offers the opportunity to shape this synthetic wavefront by setting the phase relationship between the individual beams in the array. We proposed a new technique to phase-lock on-demand and dynamically a laser array with a neural network learned in a phase correction loop. With this new iterative approach of "quasi-reinforcement learning", we demonstrated the first experimental phase locking of a large one hundred laser beams [2]. This proof of principle experiment modelled the multiple beam phase noise in a static way. The neural network (NN) only required six phase corrections to lock the array on a target phase set whatever the initial phase relationships. With this first demonstration, the neural network was learned knowing the target phase set. We present an evolution of this architecture whose learning process is now adapted to lock a laser array on any phase set adaptively. This versatile architecture is experimentally validated by locking the phases of seven amplifying beams in a multiwatts fiber laser system [3].

The learning step occurs in an optimization loop like in [2]. The NN calculates the phase corrections to apply to the multiple parallel amplifying arms from sparse intensity data collected in a speckle field, interference pattern of the beams array trough a diffuser. The new neuronal architecture is made of a first NN called "Target Adaptive Neural Network" (TANN) that provides the parameters of the second NN described in [2] depending on the target phase set. This second NN predicts the phase correction for the laser system. Its parameters can be changed and so the synthetic wavefront of the laser array, on-demand and quickly (product matrix-vector) via the TANN.

This couple of NN was implemented in a 10 kHz bandwidth feedback loop to drive the phase relationships of a fiber laser system of seven amplifying arms.

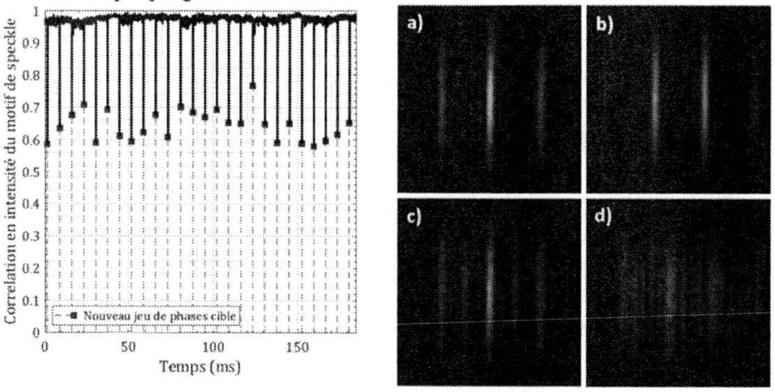

Fig. 1 Left - Experimental sequence of periodic target phase changes showing the evolution of the speckle pattern intensity correlation. Right - Examples of experimental far field patterns of phase-locked fiber laser output

The experimental study shows that the phase correction only needs six iterations to converge to the target phase set, about 1.5kHz bandwidth, with a combining efficiency of 93%. Fig. 1 shows that synthetic wavefront can be shaped at the convergence speed about each one millisecond, useful property to pre-compensate propagation distortions through aberrating medium.

References

[1] H. Tünnermann and A. Shirakawa, "Deep reinforcement learning for coherent beam combining applications," Opt. Express **27**, 24223-24230 (2019).

[2] M. Shpakovych, et al, "Experimental phase control of a 100 laser beam array with quasi-reinforcement learning of a neural network in an error reduction loop"Opt. Express **29**, 12307-12318 (2021).

[3] M. Shpakovych, et al, "On-Demand Phase Control of a 7-Fiber Amplifiers Array with Neural Network and Quasi-Reinforcement Learning"Photonics, **9**, 243, (2022)

 https://doi.org/10.1051/epjconf/202226702046

Multi-Core Fibers for Laser, Sensing and Telecommunication Applications

Thomas Schreiber[1], Stefan Kuhn[1], Johannes Nold[1], Christian Hupel[1], Sigrun Hein[1], Steffen Schulze[1], Benjamin Yildiz[1], Denny Häßner[1], Maximilian Strecker[1], Arno Klenke[2], Christopher Aleshire[2], Albrecht Steinkopf[2], Cesar Jauregui[2], Jens Limpert[1,2], Till Walbaum[1], Nicoletta Haarlammert[1]

1. Fraunhofer IOF Jena, 2. Institute of Applied Physics, FSU Jena

We will present an overview on our research on novel multi-core fibers towards lasers and amplifiers for telecommunication, sensing as well as for scaling of coherently combined high power and high energy laser systems. We will link the required properties for the fibers to the manufacturing and characterization process chain. For sensing application, the uniformity of the individual cores will be of most interest and FBG written to the cores can serve as a measure but are also required in those applications (Fig. 1). For coherently combined systems, the multi core fibers have to show additional properties, like low cross talk and high laser efficiencies (Fig. 2).

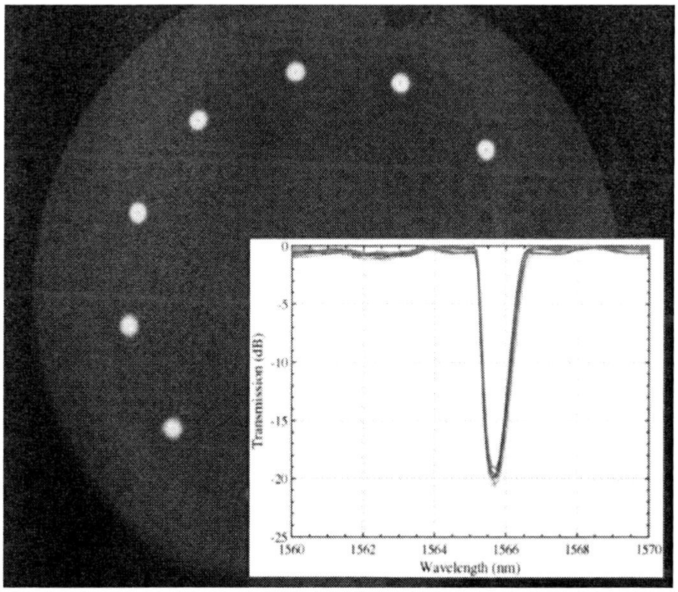

Fig. 1: Multi-Core fiber for sensing applications.

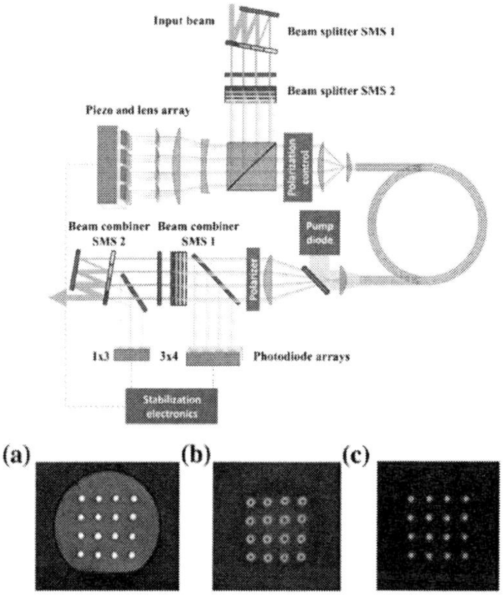

Fig. 2: Coherent combination of ultrashort laser pulses using multi core fibers.

Integrated photonic quantum systems

Michael Kues
Institute of Photonics, Leibniz University Hannover, Nienburgerstr 17, 30167, Germany

For more than 100 years, researchers explore quantum physics, which led to major breakthroughs, such as the laser, being today in daily use. With the increased control that is gained in recent years over quantum systems, the current advances in quantum technologies allow the exploitation and distinct control of quantum phenomena such as superposition and entanglement. These quantum phenomena enable fundamental performance advantages over normal classical procedures. They can be used for example to realize quantum-enhanced metrology, where quantum systems show the increase in sensitivity for metrological tasks, such as length measurements, as it will be exploited in the next generation of gravitational wave detection systems. Moreover, quantum physics is foreseen to enable quantum computing which could accelerate computational tasks for certain specific problems and enable entirely novel solutions with applications in drug development, material design or climate change. However current quantum computing technology is still in a prove-of-concept demonstration phase as it requires to prove scalability to large-scale quantum systems. More conceptually and technologically advanced is the use of quantum phenomena in order to realize future-prove communication systems, that means communication techniques which cannot be decrypted by future advances in mathematical concepts or computing resources. Here prove-of-concept systems in optical fiber as well as in satellite configurations and even commercial devices already exist, which require however advances with respect to multi-user operation increased transmission rates and reduced complexity.

Among other quantum systems, such as trapped ions or superconducting systems, photons come with the advantage that they are very robust to decoherence, i.e., they preserve the quantum state over long time and distances. Moreover, systems can mainly work at room-temperature which avoids the need for complex cryogenic cooling systems. Photons are also directly compatible with today's optical fiber communications networks that are usually used for transmitting information via classical laser pulses.

In order to realize the aforementioned applications in real-world scenarios, practical, scalable, mass-producible, system aproaches are required to transform them out of lab into industry grade systems. In this context, integrated photonic chip systems have seen tremendous advancements in recent years, and can be exploited to form stable quantum systems with the potential for scalability. In this lecture, I will give an introduction into the topic of integrated photonic quantum systems, revising the on-chip realization of quantum states and their integrated/fiber-based processing with dedicated application scenarios.

Plasmon-Empowered Nanophotonics: from Circuitry to Metasurfaces

Sergey I. Bozhevolnyi
Centre for Nano Optics, University of Southern Denmark, Campusvej 55, DK-5230 Odense M, Denmark

Surface plasmon polaritons, often shortened to surface plasmons (SPs), represent hybrid excitations involving free electron oscillations in metals and electromagnetic fields in dielectrics that propagate along and strongly bound to metal-dielectric interfaces. These surface electromagnetic waves enable deeply subwavelength confinement of guided modes along with strong enhancement of optical fields, two major features of SP modes that have been and continue being advantageously exploited in plasmon-empowered nanophotonics [1]. It would be impossible to overview, even very briefly, all fascinating topics found within plasmonics that include metasurfaces, graphene and other 2D materials, strong-coupling phenomena, topological plasmonics, quantum plasmonics, hot-electron phenomena, and many other topics even when considering only those presented at the last conference on Surface Plasmon Photonics - SPP9 (Fig 1).

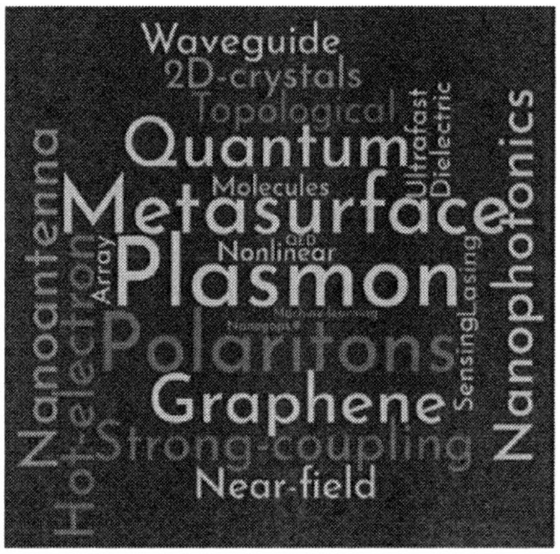

Fig. 1 The breadth of research topics and directions within plasmonics illustrated by a word cloud compiled from the book of abstract from the 9th International Conference on Surface Plasmon Photonics (SPP9) held in 2019 [2].

In this talk, special attention is given to the progress in ultra-compact photonic circuitry, including modulators and detectors, and plasmonic metasurfaces dynamically controlling propagation of light. In particular, the most efficient and ultrafast electro-optical modulators [3] and deflectors [4] utilizing the commercially viable material, LiNbO3, in which the radiation transport is controlled using the same metal circuitry for both guiding SP modes and delivering electrical signals, are presented along with the realization of on-chip electrical detection of guided SP modes [5]. Plasmonic metasurfaces [6], which can be considered as the two-dimensional analogue of metal-based metamaterials, used for room-temperature generation of single-photon streams carrying orbital angular momenta [7] and dynamic control of metasurface-enabled focusing [8, 9] and optical birefringence [10] are also discussed. A personal view on the nearest perspectives for plasmon-empowered nanophotonics concludes this talk.

References

[1] D. K. Gramotnev, and S. I. Bozhevolnyi, "Plasmonics beyond diffraction limit," Nature Photon. **4**, 83 (2010).
[2] N. A. Mortensen, P. Berini, U. Levy, and S. I. Bozhevolnyi, "Proceedings of the 9th International Conference on Surface Plasmon Photonics (SPP9) – Editorial," Nanophotonics **9**, 245 (2020).
[3] M. Thomaschewski, V. A. Zenin, C. Wolff, and S. I. Bozhevolnyi, "Plasmonic monolithic lithium niobite directional coupler switches," Nature Commun. **11**, 748 (2020).
[4] M. Thomaschewski, C. Wolff, and S. I. Bozhevolnyi, "High-speed plasmonic electro-optic beam deflectors," Nano Lett. **21**, 4051 (2021).
[5] T. Yezekyan, M. Thomaschewski, and S. I. Bozhevolnyi, "On-chip Ge photodetector efficiency enhancement by local laser-induced crystallization," Nano Lett. **21**, 7472 (2021).
[6] F. Ding, Y. Yang, R. A. Deshpande, and S. I. Bozhevolnyi, "A review of gap-surface plasmon metasurfaces: fundamentals and applications," Nanophotonics **7**, 1129 (2018).
[7] C. Wu, S. Kumar, Y. Kan, D. Komisar, Z. Wang, S. I. Bozhevolnyi, and F. Ding, "Room-temperature on-chip orbital angular momentum single-photon sources," Science Advances **8**, eabk3075 (2022).
[8] C. Meng, P. C. V. Thrane, F. Ding, J. Gjessing, M. Thomaschewski, C. Wu, C. Dirdal, and S. I. Bozhevolnyi, "Dynamic piezoelectric MEMS-based optical metasurfaces," Science Advances **7**, eabg5639 (2021).
[9]. C. Damgaard-Carstensen, M. Thomaschewski, F. Ding, and S. I. Bozhevolnyi, "Electrical tuning of Fresnel lens in reflection," ACS Photonics **8**, 1576 (2021).
[10] C. Meng, P. C. V. Thrane, F. Ding, and S. I. Bozhevolnyi, Full-range birefringence control with piezoelectric MEMS-based metasurfaces, Nature Commun. **13**, 2071 (2022).

Environmentally Stable Harmonic Modelocked All-Fibre Oscillator

Mesut Laçin[1], Paul Repgen[1], Ömer İlday[1]
Bilkent University, Ankara 06800, Turkey

Nonlinear polarisation evolution (NPE) is the most commonly used saturable absorber mechanism for fibre-based mode-locked oscillators. It provides deep modulation (50% or more), usually leads to self-starting operation, is easy to implement, and can have an all-fibre design [1]. Their inherent downside is the requirement for non-polarisation maintaining (non-PM) fibres, which makes the laser susceptible to the environmental perturbations, particularly temperature fluctuations and vibrations. This sensitivity is greatly amplified during higher-order harmonic mode-locking. We are not aware of any truly environmentally stable mechanism for passive harmonic mode-locking, which is particularly interesting as a seed source for burst-mode lasers for material processing in the ablation-cooled regime. Fundamental mode-locking is an alternative but this approach is not scalable and even at several hundred MHz, all-fibre operation based on NPE becomes increasingly difficult to self-start, typically requires specialty components, and have limited environmental stability.

Here, we present a completely fibre-integrated cavity, based on Yb-doped fibres as the gain medium. The cavity comprises only PM fibre, except for the Yb-fibre, which we coil and isolate by immersing in a heat-conductive polymer that is commonly used in double-clad fibre lasers. The setup is shown in Figure 1 (a). We have deliberately chosen the gain fibre as the only non-PM section to benefit from its isolation. The tightly coiled (to a diameter of 15 cm) gain fibre acts as a mild heat source, which is regulated by the surrounding polymer, automatically achieving passive temperature stabilization without a special effort. The polarisation can be manipulated through two fibre-based polarisation controllers at the ends of the gain fibre. The Yb-doped fibre is double-clad pumped through a PM-signal pump combiner by a wavelength-stabilized multimode pump diode at 976 nm wavelength. A fibre-integrated polarising beam splitter acts as polarisation discriminator and output coupler.

We readily obtain self-starting passive harmonic mode-locking with excellent environmental stability and short-term stability – the supermode suppression is at least 56 dB, as shown in the inset of Figure 1 (b). In the radio-frequency spectrum, the maxima at the fundamental repetition rate and lower harmonics can be efficiently suppressed, which is an important measure of the stability [2]. The intra-cavity power is estimated to be 500 mW. The output power is 250 mW, corresponding to 1 nJ pulse energy. Currently, we obtain operation at the fourth harmonic repetition rate, which corresponds to 254 MHz for this cavity. The harmonic frequency is limited only by the average power handling of the components and should be readily scalable to higher values with sufficient average power. The spectral width of the output signal is 22 nm, as shown in Figure 1 (c). The output pulse has a duration of 5.3 ps in the autocorrelation trace and can be compressed to 180 fs autocorrelation duration, and the estimated pulse width is 120 fs, assuming a Gaussian pulse shape. These parameters render the oscillator an attractive choice as a seed source in ablation-cooled material processing experiments. Further scaling of the repetition rate appears to be possible, while 250 MHz is already high enough to avoid dispersion imbalance problems in fibre-integrated repetition rate multipliers for scaling to the multi-GHz range [3].

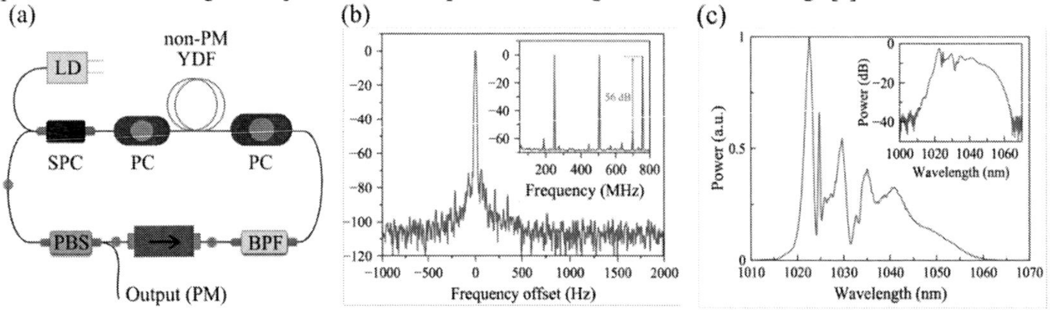

Fig. 1 Layout of the all-fibre, mostly-PM laser oscillator (a), and suppression of the supermodes in the readiofrequency spectrum (b). The output spectrum at the fourth harmonic repetition rate is shown in (c). BPF: spectral band-pass filter, LD: multi-mode pump laser diode, PBS: polarising beam splitter, PC: polarisation controller, PM: polarisation maintaining, SPC: signal-pump combiner, YDF: Ytterbium-doped fiber.

References

[1] D. Mortag, D. Wandt, U. Morgner, D. Kracht, J. Neumann, "Sub-80-fs pulses from an all-fiber-integrated dissipative-soliton laser at 1 μm" Opt. Express, **19** (2), 546 (2011).
[2] S. Zhou, D.G. Ouzounov, F.W. Wise, "Passive harmonic mode-locking of a soliton Yb fiber laser at repetition rates to 1.5 GHz", Opt. Lett., **31** (8), 1041 (2006).
[3] C. Kerse, H. Kalaycioglu, P. Elahi, Ö. Akçaalan, F.Ö. Ilday, "3.5-GHz intra-burst repetition rate ultrafast Yb-doped fiber laser", Opt. Commun., **366**, 404 (2016).

AUTHOR INDEX

AUTHOR INDEX

AUTHOR INDEX

AUTHOR INDEX

AUTHOR INDEX